T0290996

Biosecurity in the Age of Synthetic Biology

Biosecurity in the Age of Synthetic Biology is an indispensable roadmap to navigating the complex biosecurity challenges arising from the rapidly advancing field of synthetic biology. As biotechnology progresses at an unprecedented pace, the need for robust biosecurity measures has never been more critical. This book delivers meticulously crafted protocols for biosecurity risk assessments, designed to be easily followed and implemented by professionals at every level. Covering everything from physical lab security to the integration of artificial intelligence in synthetic biology research, this book offers solutions for several scenarios. In addition to technical step-by-step protocols, it thoroughly examines the ethical, legal, and societal aspects shaping the future of synthetic biology, providing a balanced view of its opportunities and inherent biosecurity risks. The author calls for strengthened biosecurity measures at every level—from local schools to international policies—to ensure the responsible development and use of synthetic biology. As the first book to offer structured guidelines for biosecurity risk assessment in synthetic biology, it is an indispensable resource for scientists, policymakers, and anyone passionate about the intersection of biotechnology and biosecurity.

Key Features:

- Step-by-step instructions and examples for biosecurity risk assessments.
- A detailed outline for creating a widely adoptable biosecurity manual.
- Insight into future challenges and opportunities in biosecurity.
- Exploration of artificial intelligence's role in synthetic biology and biosecurity.

Biosecurity in the Age of Synthetic Biology

Leyma Pérez De Haro, Ph.D.

CRC Press is an imprint of the
Taylor & Francis Group, an **informa** business

Cover: Illustration of a DNA double helix superimposed upon a circuit board background. The helix is depicted with a bioluminescent quality, glowing with a gradient from green to blue, suggesting a synergy between biological and digital information systems. The image is a metaphor for the convergence of synthetic biology and biosecurity. The author generated the image using DALL-E in ChatGPT 4.0.

Designed cover image: Leyma Pérez De Haro, Ph.D.

First edition published 2025
by CRC Press
2385 NW Executive Center Drive, Suite 320, Boca Raton FL 33431

and by CRC Press
4 Park Square, Milton Park, Abingdon, Oxon, OX14 4RN

CRC Press is an imprint of Taylor & Francis Group, LLC

This book's views and ideas are entirely the author's and not those of any past, current, or future employer or client.

Library of Congress Cataloging-in-Publication Data
Names: De Haro, Leyma Pérez, author.
Title: Biosecurity in the age of synthetic biology / Leyma Pérez De Haro.
Description: First edition. | Boca Raton : CRC Press, 2025. |
Includes bibliographical references and index.
Identifiers: LCCN 2024009689 (print) | LCCN 2024009690 (ebook) | ISBN 9781032729473 (hardback) |
ISBN 9781032725369 (paperback) | ISBN 9781003423171 (ebook)
Subjects: LCSH: Synthetic biology—Security measures. | Biosecurity.
Classification: LCC TP248.27.S95 D44 2025 (print) | LCC TP248.27.S95
(ebook) | DDC 363.325/3—dc23/eng/20240422
LC record available at https://lccn.loc.gov/2024009689
LC ebook record available at https://lccn.loc.gov/2024009690

ISBN: 978-1-032-72947-3 (hbk)
ISBN: 978-1-032-72536-9 (pbk)
ISBN: 978-1-003-42317-1 (ebk)

DOI: 10.1201/9781003423171

Typeset in Times
by codeMantra

This book is dedicated to all the biosafety and biosecurity professionals worldwide who work tirelessly every day to make this world a better place and keep it safe.

Contents

Foreword

If there's a book you want to read, but it hasn't been written yet, then you must write it.

Tony Morrison
Nobel Prize in Literature (1993)

I have known Leyma for several years and always admired her resilience, determination, grit, and perseverance. One day, she came to me and said: "*Vips, I have been asked to give a presentation about biosecurity in the context of cutting-edge developments in synthetic biology. What is the best book to read to address the risk, likelihood, and consequences within this field?*" I replied, "*I am not even sure I can think of a single textbook that meets and fits your needs, but there is an ocean of recent publications out there within both fields.*" She looked at me and said: "*Well, I can just as well write it myself!*" And so, she did. This is how I witnessed the genesis of this remarkable work, which, in my view, is set to become a cornerstone in its field.

Biosecurity in the Age of Synthetic Biology is not merely a book to be read but an experience to be engaged with, a journey to be undertaken. With great anticipation, I recommend this book to you, confident in its power to enlighten and provoke thoughtful consideration of one of the most pressing issues of our time: how to advance synthetic biology safely, securely, and responsibly.

A unique feature of this book, which I like very much, is the thought-provoking questions posed at the end of each chapter. These questions are not mere academic exercises but are intended to internalize and kick-start personal reflection. Take the time to reflect upon them. Take one question and ponder on it during your daily chores. Think deeply. These questions are a goldmine for educators responsible for preparing future generations of synthetic biologists. These educators must help students understand the potential outcomes of advancing science without caution. It's essential to consider the consequences before moving forward with experiments. Otherwise, humanity may reach a point of no return. These questions underscore the book's role not just as a source of information but as a catalyst for dialogue, urging us to ponder the "what ifs" that accompany the rapid pace of technological innovation. Furthermore, the questions are for more than just the teachers and students. They are an invitation to deep reflection, designed to engage all of us in the critical discussions that will shape our collective future. Discuss them with your colleagues, family, and friends as well.

This book focuses on the twilight zone of an evolving field moving forward at warp speed: synthetic biology and its biosecurity implications. New techniques and new methodologies emerge at an ever-increasing pace; the rate of development in the past 5 years is greater than we have seen in the past 20 years.

Nevertheless, the basic idea behind the risk identification methodologies, reflections, and systematic approaches related to the techniques, tools, and methods discussed here is universal and everlasting, and will serve the reader well in the coming years. This book is not a "one-time-read;" you'll find yourself returning several times, reflecting once more with an expanded mindset, and seeing a specific topic in a different light.

Read this book as it is intended: to start reflections and a dialogue. At first, I struggled a little with the embedded biosecurity aspect of what I normally see as biosafety. Then, I reflected deeper and realized that any biosafety aspect going wrong will often evolve into a biosecurity challenge. This book is also a road map to what the future might bring, and it is written with many different audiences in mind. The language is direct and easy to read, with as little academic slang as possible. What sets this book apart is its comprehensive examination of synthetic biology and its biosecurity implications and its profound engagement with the ethical, societal, and practical dimensions of these advancements. This book navigates these complex waters with an adept hand, merging detailed scientific analysis with broader discussions on policy, ethics, and the future trajectory of this burgeoning field. This balance renders this book an essential read for a diverse audience, from scientists and students to policymakers and the general public, anyone vested in the future of our world.

As we stand at the cusp of a new era in biological sciences, the relevance of this book cannot be overstated. It invites us to engage with the potential of synthetic biology to transform our world while confronting us with the grave responsibilities that such power entails. It is a must-read for anyone eager to participate in vital conversations about our future, offering a nuanced perspective on the challenges and opportunities ahead. You will enjoy this book as much as I did!

Vibeke (Vips) Halkjaer-Knudsen, Ph.D.

Vips is a world-renowned Biorisk Management Expert. She is the former Director of Polio Vaccine Production and former Director of EHS at the Statens Serum Institut (SSI) in Denmark.

Vips is also a retired Distinguished Member of the Technical Staff at the Center for Global Security and Cooperation at Sandia National Laboratories, United States.

Book Audience

This book will be of interest to the following groups:

Academics and Researchers: Researchers and educators in fields such as biology, bioengineering, genetics, bioinformatics, and biotechnology would find this book indispensable. It offers practical insights into biosecurity applications that align with everyday laboratory work. This alignment with real-world scenarios enriches the educational experience and research practices, fostering a safety and ethical responsibility culture.

Biosafety and Biosecurity Professionals: This book serves as a comprehensive guide for professionals dedicated to biosafety and biosecurity. It provides a detailed discussion of the risks associated with synthetic biology and offers practical guidelines for ensuring biosafety. Specifically, this book details the methodologies for conducting a biosecurity risk assessment at their institution. This structured approach enhances the efficiency and effectiveness of risk management, ensuring a robust defense against potential biosecurity threats.

Industry Professionals: Individuals engaged in sectors influenced by synthetic biology, including biotechnology, pharmaceuticals, and agriculture, would find this book an essential resource. They can safeguard intellectual property and technological assets by staying informed about the current developments, challenges, and biosecurity measures, thereby reducing economic espionage and theft of intellectual property. This in-depth understanding facilitates informed decision-making and fosters a proactive approach to risk management, contributing to industry stability and growth.

Policymakers and Regulators: For those entrusted with shaping and enforcing policies and regulations related to biosafety, biosecurity, and synthetic biology, this book offers a well-articulated framework. It can be utilized as a foundational resource to guide their work in creating a global manual for biosecurity, leveraging the levels and strategies proposed within. This alignment with a meticulously researched guide ensures that policies and regulations are grounded in evidence-based practices, enhancing their credibility and effectiveness. It supports the collaborative effort to establish international standards, promoting a unified approach to biosecurity that reflects the interconnected nature of today's global scientific community.

This book's multifaceted approach and real-world applicability make it an invaluable tool for these diverse groups, addressing their unique needs and contributing to the broader goal of advancing biosecurity in the age of synthetic biology. Its relevance, practicality, and timely insights underscore its potential as a vital resource for shaping the future of science, industry, and policy.

Preface

Thank you for choosing to engage with *Biosecurity in the Age of Synthetic Biology*. This book is designed to help investigate the intersection of synthetic biology and biosecurity, delving into the potential risks and rewards of this rapidly advancing field. Born from a desire to fill in an identified gap as a way to contribute to the ongoing discourse and provide a comprehensive resource, this project has been both challenging and deeply rewarding. This book is structured into 13 chapters, each focusing on a unique aspect of synthetic biology and its implications for biosecurity.

The Author's Journey: As a researcher in the field, I have long been captivated by the potential synthetic biology holds to revolutionize our world, specifically biomedicine, over the next decade. I recognize the need for stringent biosecurity measures to ensure the responsible development and deployment of these technologies for good, while minimizing the risks of negative consequences either accidental or intentional. This book is the result of a passion for these subjects, fueled by countless hours of research and exploration. With expertise in biorisk management and synthetic biology, I have endeavored to bring a unique perspective to this text, and I hope it will help leap forward the field to keep up with the technology.

The Relevance of the Topic: Synthetic biology holds immense promise, but it also poses significant biosecurity challenges. This book addresses a critical gap in the existing literature by examining these issues in depth. It is my hope that this work will contribute to a greater understanding of these challenges, fostering dialogue and action to ensure a safe and secure future in the age of synthetic biology.

Notes on Approach and Scope: This book takes a broad approach, covering a range of topics related to synthetic biology and biosecurity. However, it is by no means exhaustive. The field is rapidly evolving, and new challenges and opportunities are continually emerging.

How to Use This Book: The chapters are designed to be self-contained but interconnected, allowing readers to explore topics of particular interest or to read the book from start to finish. I encourage readers to seek out additional resources and stay informed about new developments in the field.

Invitation to Engage: I invite you, the reader, to actively engage with the material in this book. Reflection, discussion, and critical thinking are encouraged. I look forward to hearing your thoughts and perspectives on these important issues on a dedicated page (https://www.linkedin.com/company/biosecurity-synbio/).

Finally, thank you again for embarking on this journey with me. I hope that this book will inspire you to reflect on the profound implications of synthetic biology for our world and contribute to the ongoing dialogue about how to ensure a safe and secure future. Let us face these challenges together and navigate toward a promising and secure future.

About the Author

Dr. Leyma Pérez De Haro earned a B.S. in Biochemistry from California State University, Los Angeles, and holds a Ph.D. in Biomedical Sciences from the University of New Mexico. She is a Registered Biosafety Professional (RBP) by ABSA International. She completed two postdoctoral fellowships, including one at Los Alamos National Laboratory. Dr. De Haro worked at Sandia National Laboratories, in the internationally renowned Global Chemical and Biological Security Team, where she specialized in Biosafety and Biosecurity on a global scale. She is currently part of the Biosafety Team at the California Institute of Technology. Dr. De Haro has over 14 years post-Ph.D. experience as a scientist, helping laboratories and organizations enhance their biosafety and biosecurity practices. Her background in innovative scientific research enables her to understand the unique safety challenges that laboratories and organizations face in today's rapidly evolving world of life sciences. With a keen focus on fostering a culture of responsibility and safety, Dr. De Haro is committed to promoting best practices in biosafety and biosecurity worldwide, supporting the life sciences community in pursuing groundbreaking discoveries and innovations.

Acknowledgments

I am deeply grateful to all those who contributed to the creation of this book. The knowledge and insight shared by numerous experts and mentors have been invaluable. Specifically, I would like to extend my sincere gratitude to the following individuals for their invaluable assistance and expertise in revising and enhancing the content of this book: Vibeke Halkjaer-Knudsen, Ph.D., for her expertise and advice in the field of Biosecurity and Biorisk Management. Sandra D. Mied, B.A., for her expertise in counterintelligence and her thoughtful review and edits. Reza Sadri, Ph.D., for his expertise and advice in the field of AI. My editor C. R. "Chuck" Crumly, Ph.D., for his enthusiasm and support for this project, but most importantly for believing in me. Last but not least, Drago for believing in me and supporting me throughout this project. Their thoughtful feedback and constructive suggestions significantly enhanced the quality and clarity of this work.

List of Abbreviations

ABAI	Agricultural Biosecurity Authority of India
ADHD	Attention Deficit Hyperactivity Disorder
AI	Artificial Intelligence
AIA	Advanced Informed Agreement
ANLIS-Malbrán	Antimicrobial Service of the National Institute of Infectious Diseases – Dr. Carlos G. Malbrán
APHIS	Animal and Plant Health Inspection Service
AU	African Union
AU-BBI	African Union Biosafety and Biosecurity Initiative
BCH	Biosafety Clearinghouse
BLAST	Basic Local Alignment Search Tool
BMBL	Biosafety in Microbiological and Microbiology Laboratories
BMENA	Broader Middle East and North Africa
BPR	Biodefense Posture Review
BRM	Biorisk Management
BRO	Biorisk Management Officer
BSC	Biological Safety Cabinet
BSec-L	Biosecurity Levels
BSL	Biosafety Levels
BSO	Biosafety/Biosecurity Officer
BTWC	Biological and Toxin Weapons Convention
BW	Biological Weapon
BWC	Biological Weapons Convention
C. diff	*Clostridium difficile*
CARs	Chimeric Antigen Receptors
CAR-T	Chimeric Antigen Receptor T-Cell Therapy
CBD	Convention on Biological Diversity
CBRN	Chemical, Biological, Radiological, and Nuclear
CBS	Canadian Biosafety Standard
CDC	Centers for Disease Control and Prevention
CEN	Comité Européen de Normalisation (French) or European Committee for Standardization (English)
CHIPS	Creating Helpful Incentives to Produce Semiconductors for America
CONICET	National Commission for the Conservation and Sustainable Use of Biodiversity
COSPAR	Committee on Space Research
COVID-19	Coronavirus Disease 2019
CRISPR	Clustered Regularly Interspaced Short Palindromic Repeats
CWA	CEN Workshop Agreement
DETECTR	DNA Endonuclease-Targeted CRISPR Trans Reporter
DIY	Do-It-Yourself

DURC	Dual-Use Research of Concern
EBN	European Biosafety Network
EBSA	European Biosafety Association
EPA	Environmental Protection Agency
EPO	Erythropoietin
FAO	Food and Agriculture Organization
FBI	Federal Bureau of Investigation
FDA	Food and Drug Administration
FMD	Foot and Mouth Disease
FOXO3	Forkhead Box O3
FRAND	Fair, Reasonable, and Non-Discriminatory
GIBACHT	Global Partnership Initiated Biosecurity Academia for Controlling Health Threats
GM	Genetically Modified
GMO	Genetically Modified Organism
GMPP	Good Microbiological Practice and Procedure
HBV	Hepatitis B Virus
HEXA	Hexosaminidase A Gene
HFEA	Human Fertilization and Embryology Authority
hGH	Human Growth Hormone
HIV	Human Immunodeficiency Virus
HVAC	Heating, Ventilation, and Air Conditioning
IBC	Institutional Biorisk Management Committee
ICAR	Indian Council of Agricultural Research
ID50	Infectious Dose for 50% of the Population
iGEM	International Genetically Engineered Machine
IGSC	International Gene Synthesis Consortium
IP	Intellectual Property
IRB	Institutional Review Board
ISO	International Organization for Standardization
IT	Information Technology
IVF	*In Vitro* Fertilization
LCA	Leber's Congenital Amaurosis
LMOs	Living-Modified Organisms
MEXT	Ministry of Education, Culture, Sports, Science, and Technology (Japan)
MIT	Massachusetts Institute of Technology
Mpox	Virus formerly known as monkeypox
mRNA	Messenger Ribonucleic Acid
NDA	Non-Disclosure Agreement
NGS	Next-Generation Sequencing
NIH	National Institutes of Health
PAPR	Powered Air-Purifying Respirator
PCR	Polymerase Chain Reaction
PCSBI	Presidential Commission for the Study of Bioethical Issues
PGD	Pre-implantation Genetic Diagnosis

PI	Principal Investigator
PPE	Personal Protective Equipment
PHAC	Public Health Agency of Canada
REGN-COV2	Regeneron's Antibody Cocktail
RFID	Radio Frequency Identification
RSV	Respiratory Syncytial Virus
RT-PCR	Reverse Transcription Polymerase Chain Reaction
SA	Select Agents
SARS-CoV-2	Severe Acute Respiratory Syndrome Coronavirus 2
SB3	Brazilian Biosafety & Biosecurity Society
SDS	Safety Data Sheet
SEAS	Southeast Asia Strategic Biosecurity Dialogue
SHERLOCK	Specific High-sensitivity Enzymatic Reporter unLOCKing
SOP	Standard Operating Procedure
TALENs	Transcription Activator-Like Effector Nucleases
UNICRI	United Nations Interregional Crime and Justice Research Institute
USAMRIID	U.S. Army Medical Research Institute of Infectious Diseases
USDA	United States Department of Agriculture
UV	Ultraviolet
VBM	Valuable Biological Material
VLP	Virus-Like Particle
WGS	Whole Genome Sequencing
WHO	World Health Organization
WOAH	World Organization for Animal Health
XNA	Xeno Nucleic Acid
ZFNs	Zinc Finger Nucleases

Glossary

Accident: An unintended event that causes harm, such as infection, illness, injury in humans, or contamination of the environment.

Antibody: An antibody, also known as immunoglobulin, is a specialized protein produced by the immune system in response to the presence of specific antigens. Antibodies are created by B lymphocytes, a type of white blood cell, and are designed to recognize and bind to specific antigens. They play a crucial role in the immune response by recognizing foreign or harmful substances, such as microorganisms (bacteria, viruses, fungi), toxins, or even abnormal cells, and neutralizing or eliminating them. Antibodies act by targeting and binding to antigens, either directly neutralizing their activity, marking them for destruction by other immune cells, or activating other components of the immune system to mount a response. The production of antibodies is a fundamental aspect of the adaptive immune response, providing the immune system with a means to specifically recognize and combat a wide range of antigens.

Antigen: An antigen is a substance, typically a protein or a molecule, which can elicit an immune response in the body. Antigens are recognized by the immune system as foreign or non-self, and they stimulate the immune system to produce an immune response, including the production of antibodies. Antigens can be derived from various sources, such as pathogens (bacteria, viruses, parasites), toxins, foreign tissues (in the case of transplantation), or environmental substances (such as pollen or certain chemicals). When the immune system encounters an antigen, it triggers a series of immune reactions to eliminate or neutralize the perceived threat. The recognition and response to antigens play a crucial role in the immune system's ability to protect the body from infections and other harmful substances.

Artificial Intelligence: Artificial Intelligence (AI) is a branch of computer science dedicated to creating systems capable of performing tasks that typically require human intelligence. These tasks include, but are not limited to, learning from data (machine learning), understanding natural language (natural language processing), recognizing patterns and images (computer vision), and making decisions based on complex datasets (decision-making algorithms). AI technologies aim to simulate aspects of human cognition and can range from specific applications, such as speech recognition and chatbots, to more complex systems that can autonomously perform a variety of tasks.

Autoclave: This is a machine used to sterilize equipment and other objects using high-pressure steam. This kills bacteria, viruses, fungi, and spores.

Bacteria: These are single-celled organisms that can live in diverse environments, including extreme temperatures and pressures, radioactive waste, and the human body. They play key roles in the Earth's ecosystem, and some can cause disease in humans.

BioBrick: A standardized biological building block used in synthetic biology to engineer and construct genetic systems. A BioBrick consists of a specific DNA sequence that encodes a functional genetic element, such as a gene or a regulatory element and is designed to be easily interchangeable and combinable with other BioBricks. These modular components allow researchers to assemble and reconfigure genetic circuits to create novel organisms and biological functions. BioBrick standardization enables efficient collaboration and promotes the rapid development of biotechnological applications.

Biohazard: This is a biological substance that threatens the health of living organisms, primarily humans. This could include a sample of a microorganism, virus, or toxin that can affect human health.

Biohazardous waste: This is any waste containing infectious materials or potentially infectious substances such as blood. Of special concern are sharp wastes such as needles, blades, glass pipettes, and other wastes that can cause injury during handling.

Biological agent: This is a microorganism, virus, biological toxin, particle, or any infectious material that could potentially cause infection, allergy, toxicity, or other hazards to humans, animals, or plants. It can be natural or genetically modified.

Biological containment: This is the use of biological measures to control the spread of potentially harmful biological agents. This can include the use of vaccines or other medical treatments to prevent infection, or the use of genetically modified organisms that are unable to survive outside the laboratory.

Biological exposure: This is the contact a person has with a biological agent that could result in harm, such as an infectious disease or toxic reaction. The exposure can occur through different routes, such as inhalation, ingestion, or direct contact with the skin or eyes.

Biological hazard: This is a potential source of harm caused by a biological agent or condition. Examples include infectious diseases, exposure to harmful biological toxins, or contact with biohazardous materials.

Biological risk: This is the risk associated with exposure to biological agents, such as bacteria, viruses, or toxins, which can harm human health.

Biological safety cabinet (BSC): This is a closed, ventilated workspace designed to protect the user, the laboratory environment, and/or the work materials from aerosol hazards. It works by separating the work from the rest of the lab and controlling the airflow. The air that leaves the cabinet is filtered to remove microbes before it enters the lab or the building's ventilation system. There are different classes of BSCs (I, II, and III) that provide different levels of protection.

Biological safety: This refers to the practices and procedures followed to prevent exposure to infectious agents or biohazards.

Biological warfare: This is the use of biological toxins or infectious agents such as bacteria, viruses, insects, and fungi with the intent to kill or incapacitate humans, animals, or plants as an act of war.

Biosafety committee: This is a group within an institution that independently reviews biosafety issues and reports to senior management. The committee should include representatives from different areas of the organization who have scientific expertise.

Biosafety level: This is a set of biocontainment precautions required to isolate dangerous biological agents. The levels of containment range from the lowest biosafety level 1 (BSL-1) to the highest at level 4 (BSL-4). The higher the number, the higher the containment level, and the more dangerous the biological agents.

Biosafety officer: This is a person assigned to oversee the biosafety (and biosecurity) programs of a facility or organization. This person may have different titles such as biosafety professional, biosafety advisor, biosafety manager, biosafety coordinator, or biosafety management advisor.

Biosafety program management: This involves developing, implementing, and overseeing biosafety at the organizational level. It includes creating policies, guidance documents for practices and procedures, planning documents (for training, recruitment, and emergency/incident response), and keeping records (personnel, inventories, and incident management).

Biosafety: This is a field that uses principles, technologies, and practices to prevent accidental exposure to biological agents or their unintentional release.

Biosecurity: This refers to the principles, technologies, and practices implemented to protect and control biological materials and/or the equipment, skills, and data related to their handling. The goal is to prevent unauthorized access, loss, theft, misuse, diversion, or release.

Certification: This is a formal confirmation by a third party that a system, person, or piece of equipment meets certain requirements, often according to a certain standard. It is a way of assuring that something is up to par.

CHIPS and Science Act: A 2022 U.S. federal statute that provided new funding to boost domestic research and manufacturing of semiconductors. The act intends to ensure that the United States retains leadership in the semiconductor industry.

Communicability: This is the ability of a biological agent to be transmitted from one person or animal to another, either through direct or indirect transmission. It is often related to an epidemiological measurement called the basic reproduction number (R_0), which is the average number of new infections caused by a single infected individual in a population where everyone is susceptible.

Consequence (of a laboratory incident): This is the outcome of an incident, such as exposure to or release of a biological agent, of varying severity of harm, which occurs during laboratory operations. Consequences may include a laboratory-associated infection, other illness or physical injury, environmental contamination, or asymptomatic carriage of a biological agent.

Contagious: This refers to a disease spread from one person or organism to another by direct or indirect contact. Contagious diseases can be spread in several ways, including inhaling airborne germs, touching contaminated surfaces, or through skin-to-skin contact.

Containment: This refers to the combination of physical structures and operational practices that protect people, the work environment, and the community from exposure to biological agents. "Biocontainment" is also used in this context.

Core requirements: These are the basic requirements defined in the fourth edition of the World Health Organization (WHO) Laboratory biosafety manual to describe a combination of measures that form the basis of laboratory biosafety. They reflect international standards and best practices in biosafety that are necessary for safe work with biological agents.

Decontamination: Decontamination refers to a systematic process that aims to remove, neutralize, or reduce contaminants, including microorganisms, hazardous chemicals, radioactive substances, or infectious agents, from objects, surfaces, or environments. The process involves the use of chemical or physical means to eliminate or decrease the presence of these contaminants to a predetermined level of safety. Decontamination procedures can vary depending on the nature of the contaminants and the specific object or surface being treated. Chemical decontamination agents, such as disinfectants or specialized cleaning solutions, are often used to neutralize or kill microorganisms. Physical methods, such as heat, pressure, or filtration, may also be employed to remove or reduce contaminants. Decontamination is a critical practice in various settings, including healthcare facilities, laboratories, hazardous material handling, and emergency response, as it helps to minimize the risk of infection, ensure workplace safety, and protect public health.

Disease surveillance: Disease surveillance is a continuous and systematic process that involves the collection, analysis, interpretation, and dissemination of health-related data to monitor and track the occurrence, distribution, and trends of diseases within a population or specific geographic area. The primary objectives of disease surveillance are early detection, timely response, and prevention of disease outbreaks or other public health events. Surveillance systems collect data from various sources, such as healthcare facilities, laboratories, and community reports, and analyze the information to identify patterns, trends, and potential threats to public health. By monitoring disease patterns, surveillance enables public health authorities to identify and respond to outbreaks, assess the effectiveness of control measures, and guide public health interventions and policies. Disease surveillance is an essential tool for maintaining public health and plays a critical role in preventing and controlling the spread of infectious diseases.

Disinfection: Disinfection is a systematic process that involves the elimination or reduction of viable microorganisms, such as bacteria, viruses, and fungi, from surfaces or objects. It is aimed at lowering the risk of infection by destroying or inactivating pathogens present on these surfaces. Disinfection can be achieved using various chemical agents or physical methods, depending on the nature of the item or surface being treated. Chemical disinfectants, such as bleach, alcohol, or hydrogen peroxide, are commonly used to kill or inactivate microorganisms. Physical disinfection methods may include heat, ultraviolet (UV) radiation, or filtration. The effectiveness of disinfection is influenced by factors such as contact time, concentration or intensity of the disinfectant, and the specific microorganisms targeted. Disinfection plays a crucial role in infection prevention and control, particularly in healthcare settings, laboratories, and public areas where the risk of pathogen transmission is a concern.

Dual-use research: This refers to materials, information, and technologies that are intended for benefit, but could potentially be misused to cause harm. A classic example of this is a kitchen knife that could be considered dual use because it is used to cut vegetables and other food items in the context of cooking, but it could also be used as a weapon by a person with ill intent. An example of a modern technology that can be used for both beneficial and harmful purposes is "drones." For example, drones can be used to deliver packages, discover archeological sites, inspect bridges and other infrastructure, search for lost people, and monitor wildlife. However, drones can also be used for malicious purposes, such as delivering explosives or other weapons, conducting surveillance, spying on people, spreading disinformation, or even interfering with communications. As a result, drones are a dual-use technology.

Emergency/incident response: This is a planned set of actions to respond to unexpected situations, including exposure to or release of biological agents. The goal is to prevent injuries or infections, reduce damage to equipment or the environment, and speed up the return to normal operations.

Emerging infectious disease: This refers to a newly identified or reemerging disease rapidly increasing in incidence or geographic range. Emerging infectious diseases pose a public health threat and require close monitoring and response.

Environmental monitoring: This is the systematic collection and analysis of data to assess the presence and levels of biological agents or hazardous substances in the environment. It helps identify potential risks and supports the implementation of appropriate control measures.

Exposure control plan: This plan identifies potential exposures to biohazards in the workplace and outlines procedures, practices, and personal protective equipment designed to reduce the risk of exposure.

Exposure: This is an event during which a person comes into contact with, or is in close proximity to, biological agents with the potential to cause infection or harm. This can happen through inhalation, ingestion, injury, or absorption. Some infection routes are specific to the laboratory environment and are not commonly seen in the general community.

Gain-of-function: Gain-of-function refers to a type of research that involves the manipulation of an organism's genes to enhance or introduce new functionalities. Specifically, this approach aims to increase the transmissibility, host range, or virulence of a pathogen. Gain-of-function research is often undertaken to understand disease mechanisms, identify potential vulnerabilities, and develop medical countermeasures. However, it also raises biosafety and biosecurity concerns due to the potential for unintended release or misuse.

Genetic engineering: This is the process of manipulating the genes of an organism to introduce desired traits or characteristics. It involves altering the

DNA sequence to achieve specific outcomes, such as producing proteins or enhancing disease resistance.

Genomic sequencing: Genomic sequencing is the process of determining the complete DNA sequence of an organism, such as a virus or bacteria. It provides detailed information about the genetic makeup and evolution of the organism.

Global health security: Global health security refers to the collective efforts and measures taken to prevent, detect, and respond to infectious disease threats that could have an international impact. It involves collaboration between countries, organizations, and stakeholders to strengthen health systems and preparedness.

Good microbiological practice and procedure (GMPP): This is a basic code of practice for all types of laboratory activities with biological agents. It includes general behaviors and aseptic techniques that should always be observed in the laboratory to protect laboratory personnel and the community from infection, prevent contamination of the environment, and provide protection for the work materials in use.

Hand hygiene: Hand hygiene refers to cleaning hands to reduce the risk of transmitting infectious agents. It includes washing hands with soap and water or using hand sanitizers with at least 60% alcohol content.

Hazard: This is something that has the potential to cause harm when an organism, system, or population is exposed to it. In laboratory biosafety, the hazard is biological agents which have the potential to cause adverse effects to personnel and/or humans, animals, and the wider community and environment.

Heightened control measures: These are a set of control measures described in the WHO Laboratory biosafety manual that may need to be applied in a laboratory facility because the risk assessment indicates that the biological agents being handled and/or the activities to be performed with them are associated with a risk that cannot be brought below an acceptable level with the core requirements only.

Host: A host is an organism (human, animal, or plant) that harbors and provides a living environment for a pathogen.

Host range: This refers to the range of host species that a particular pathogen can infect and potentially cause disease in. Different pathogens have different host range specificities.

Immune response: This is the collective response of the immune system to the presence of an antigen, such as a pathogen. The immune response involves various cells and molecules that work together to recognize, neutralize, and eliminate the antigen.

Immunity: Immunity refers to the state of protection or resistance exhibited by an organism against a particular disease or infectious agent. The immune system can recognize and effectively respond to pathogens, preventing or limiting infection and disease. Immunity can be acquired through natural means, such as previous exposure to and recovery from an infection, which stimulates the production of specific antibodies or immune cells. Additionally, immunity can be acquired artificially through vaccination, where a vaccine stimulates the immune system to generate a protective response without causing the disease itself. Immunity plays a vital role in maintaining the overall health and well-being of individuals and populations by reducing the likelihood of infection, limiting the severity of disease if infection occurs, and contributing to herd immunity within communities.

Inactivation: This is the process of removing the activity of biological agents by destroying or inhibiting their ability to reproduce or carry out enzymatic activity.

Incidence: In epidemiology, incidence refers to the number of new cases of a specific disease or condition that occur within a defined population over a specified time period. It provides information about the risk of developing the disease.

Incident: This is an event that has the potential to, or results in, the exposure of laboratory personnel to biological agents and/or their release into the environment that may or may not lead to actual harm.

Incubation period: The incubation period refers to the duration between the time an individual is exposed to an infectious agent and the onset of the first symptoms or signs of the disease. It represents the time it takes for the pathogen to establish an infection, replicate within the body, and reach a sufficient level to produce noticeable clinical manifestations. The length of the incubation period can vary significantly depending on factors such as the specific infectious agent, the route of infection, the individual's immune response, and other host-related factors. Understanding the incubation period is essential for disease surveillance, diagnosis, and implementing appropriate public health measures to control the spread of the disease.

Infectious dose: This is the amount of a biological agent needed to cause an infection in the host, measured in the number of organisms. It is often defined as the ID_{50}, the dose that will cause infection in 50% of those exposed.

Laboratory biosafety: Laboratory biosafety refers to the practices, procedures, and containment measures implemented to prevent the accidental exposure and/or release of infectious agents or toxins from laboratories. It ensures the safety of laboratory personnel and the surrounding environment.

Likelihood (of a laboratory incident): This is the probability that an incident (such as exposure to or a release of a biological agent) will occur during laboratory work.

Maximum containment measures: These are a set of control measures described in the WHO Laboratory biosafety manual that may need to be applied in a laboratory facility because the risk assessment indicates that the biological agents being handled and/or the activities to be performed with them are associated with a risk that cannot be brought below an acceptable level with the heightened control measures only. These are a set of highly detailed and stringent control measures that are necessary for laboratory work where the activities performed pose the highest risks to laboratory personnel, the wider community, and/or the environment, and therefore, a maximum level of protection must be provided.

Mitigation: This refers to actions and measures taken to reduce or minimize the impact of a hazard or risk, such as an infectious disease outbreak. Mitigation strategies can include preventive measures, control measures, and interventions to reduce transmission.

Mutation: This is a change in the DNA sequence of a gene or a genome. Mutations can occur naturally or can be induced by various factors, and they can lead to genetic variations and diversity.

One Health: This approach involves multiple sectors communicating and working together to achieve better public health outcomes. It is particularly relevant in areas like food safety, controlling diseases that can be passed between animals and humans (zoonoses), and fighting antibiotic resistance.

Outbreak: This refers to the occurrence of cases of a particular disease in a population or geographic area that is greater than what is typically expected. Outbreaks can vary in size and severity, from localized to widespread and mild to severe.

Pathogen: A pathogen is a microorganism, such as a bacterium, virus, fungus, or parasite, which can cause disease in a host organism. Pathogens can infect and replicate within the host, leading to illness or infection.

Pathogenicity: This is the ability of an organism, a pathogen, to cause a disease in a host organism. It is a measure of how good a pathogen is at causing disease.

PCR (Polymerase chain reaction): This laboratory technique is used to amplify a specific segment of DNA or RNA. PCR allows scientists to detect and analyze genetic material, including that of pathogens.

Personal protective equipment (PPE): Personal protective equipment (PPE) encompasses a range of specialized clothing, masks, gloves, and equipment designed to protect individuals from exposure to hazardous substances, infectious agents, or other risks. PPE is a crucial component in ensuring the safety and well-being of individuals in various settings, including healthcare, laboratories, and hazardous work environments. The specific items included in PPE may vary depending on the nature of the hazards present. PPE is particularly important for preventing transmission and minimizing the risk of infection. It is a barrier between the wearer and potentially harmful agents, reducing the likelihood of exposure and contamination. The selection and use of appropriate PPE are guided by the specific hazards and risks involved to ensure effective protection and mitigate potential health and safety risks.

Prevalence: In epidemiology, prevalence refers to the total number of cases of a disease or condition present in a population at a given time. It provides information about the burden of the disease in the population.

Procedure: This refers to a series of actions conducted in a certain order or manner. In a lab, this could refer to how a particular experiment is performed, how equipment is cleaned, or how samples are stored.

Resilience: Resilience refers to the ability of individuals, communities, and systems to adapt, recover, and thrive in the face of adversity or challenging circumstances. It involves building capacity and resources to withstand and respond to health threats.

Risk: This is the chance that something harmful will happen. In a lab, this often refers to the chance that someone will be exposed to a biological agent or that the agent will be released into the environment.

Risk assessment: Risk assessment is a systematic and structured process used to evaluate potential risks associated with a specific hazard or situation. It involves identifying and analyzing hazards, estimating the likelihood of those hazards causing harm, and assessing the potential severity of the harm that could occur. Risk assessments consider various factors such as the nature of the hazard, exposure pathways, and the vulnerability of individuals or populations involved. By considering these factors, risk assessments provide valuable information for decision-making and help prioritize appropriate risk management strategies to mitigate or control the identified risks effectively.

Risk communication: Risk communication is a multifaceted process that involves the exchange of information regarding risks, potential hazards, and appropriate actions between experts, decision-makers, and the public. It is an essential component of public health and safety, particularly in managing risks associated with infectious diseases. Effective risk communication entails the clear, timely, and understandable dissemination of information to individuals and communities, tailored to their specific needs and understanding. The objective of risk communication is to promote understanding, informed decision-making, and the adoption of appropriate measures to manage risks effectively. It emphasizes transparency, accessibility, and the engagement of various stakeholders to facilitate a shared understanding of risks and encourage collective action.

Risk control: This is the process of implementing measures to reduce the likelihood of a risk event occurring or to minimize the impact if it does occur. It could involve things like using safer procedures, using personal protective equipment, or improving training.

Risk management: This is the process of identifying, assessing, and controlling risks. It involves figuring out what could go wrong, how likely it is, how bad it could be, and then deciding what steps to take to prevent it.

Risk stratification: This is the process of classifying individuals or populations into different risk categories based on their likelihood of developing a particular disease or experiencing adverse outcomes. Risk stratification helps inform targeted interventions and resource allocation.

Route of transmission: This is how an infectious agent is passed from its reservoir to a susceptible host. Typical routes include air, direct contact, indirect contact, contaminated food or water, and vector-borne transmission.

Safety equipment: These are devices used to protect the health and safety of workers. In a lab, this could include things like fume hoods, biological safety cabinets, and fire extinguishers.

Safety measures: These are steps to reduce the risk of accidents and injuries. In a lab, this could include wearing personal protective equipment, following safe work procedures, and using safety equipment.

Select agents: These are biological agents and toxins that have been identified for regulation because they potentially pose a severe threat to public health and safety. They could be used in biological warfare or bioterrorism, so labs that work with these agents must follow strict regulations.

Specificity: In diagnostic tests, specificity is the ability of a test to correctly identify individuals who do not have the disease (true negatives). It is calculated as the proportion of true negatives among all disease-free individuals.

Spore: This is a tough, protective shell form that some bacteria and fungi produce to survive harsh conditions. Spores can survive without nutrients and are resistant to heat, desiccation, and disinfectants.

Standard operating procedures (SOPs): These are detailed, written instructions on how to perform a specific task or operation that is conducted in a certain way. SOPs are used to promote work consistency and efficiency and to ensure quality and compliance with regulations.

Standard precautions: Standard precautions are basic infection prevention practices that should be followed in all healthcare settings to minimize the risk of infection transmission. They include hand hygiene, personal protective equipment, safe injection practices, and respiratory hygiene.

Sterilization: This is a process, either physical or chemical, that kills or inactivates all forms of microbial life, including hard-to-kill bacterial spores.

Surveillance: Surveillance is the ongoing systematic collection, analysis, and interpretation of health-related data. This information is then used for planning, implementing, and evaluating public health interventions and programs.

Surveillance system: A surveillance system refers to a structured and organized approach that involves the collection, analysis, and interpretation of health-related data to monitor, detect, and track patterns or trends in disease occurrence within a population. These systems play a crucial role in public health by providing timely and relevant information for monitoring the overall health status, detecting and investigating outbreaks, and guiding appropriate public health interventions. Surveillance systems are designed to facilitate the early detection and response to potential public health threats, enabling proactive measures to mitigate the impact of diseases on individuals and communities.

Toxin: This is a poison produced by living cells or organisms. Some bacteria, plants, and animals produce toxins that can be harmful or lethal to other organisms.

Transmission: This is the passing of a disease-causing agent (pathogen) from an infected host individual or group to a particular individual or group, regardless of whether the other individual was previously infected.

Vaccine: This is a biological preparation that provides active acquired immunity to a particular infectious disease. A vaccine typically contains an agent that resembles a disease-causing microorganism and is often made from weakened or killed forms of the microbe, its toxins, or one of its surface proteins.

Validation: This is proving that a method works as expected and provides evidence that the performance meets the requirements for a specific use.

Valuable biological materials (VBM): VBM according to the WHO may include pathogens and toxins, as well as non-pathogenic organisms, vaccine strains, foods, genetically modified organisms (GMOs), cell components, genetic elements, and extraterrestrial samples.

Vector: In epidemiology, a vector is any agent (animal, insect, or microorganism) that carries and transmits an infectious pathogen into another living organism.

Verification: This is the process of showing that a method, which has already been validated (proven to work), still works as expected when used by the user. It is a way of checking that everything is still working correctly.

Viable: In terms of microbiology, viable refers to a cell or organism that is capable of living or growing. For example, a viable bacterium can divide and form a colony.

Viral load: This is the amount of virus in an organism's body. It is often used for HIV and hepatitis C, but can be used for any virus. It is usually measured in the blood but also in other body fluids.

Virulence: This measures the severity of a disease caused by a microorganism. The more virulent a pathogen, the more severe the disease it causes.

Virus: This is a type of microorganism that is smaller than bacteria and cannot grow or reproduce apart from a living cell. A virus invades a living cell and uses the cell's chemical machinery to keep itself alive and to replicate itself.

Zoonosis (plural: zoonoses): This is a disease or infection that can be naturally transmitted between animals and humans.

Zoonotic disease: A zoonotic disease, also known as zoonosis, refers to a type of disease that can be transmitted between animals and humans. It occurs when a disease-causing agent, such as a virus or bacteria, can jump from an animal host to humans through direct or indirect contact. Some common examples of zoonotic diseases include rabies, Lyme disease, and certain forms of influenza.

1 The Case for Considering Biosecurity in Synthetic Biology

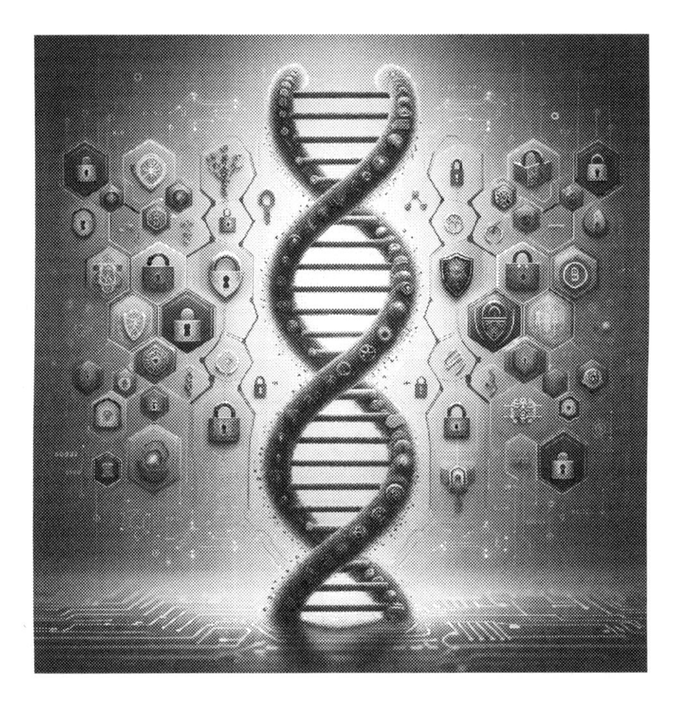

This chapter presents a compelling argument for integrating biosecurity measures into synthetic biology. It highlights this rapidly evolving discipline's unique challenges and potential risks. It underscores the importance of biosecurity in preventing the misuse of synthetic biology for harmful purposes, such as bioterrorism or the accidental release of hazardous biological agents. The chapter discusses the biosecurity risks inherent in synthetic biology, illustrated by specific real-world cases and lessons learned. The chapter presents a forward-looking perspective, emphasizing the need for ongoing vigilance and adaptive strategies to ensure biosecurity keeps pace with the advancements in synthetic biology. It calls for a collaborative effort involving scientists, policymakers, ethicists, and security experts to safeguard against the potential misuse of synthetic biology while promoting its positive applications for the betterment of society.

DOI: 10.1201/9781003423171-1

1.1 OVERVIEW OF SYNTHETIC BIOLOGY

Synthetic biology is a field that merges biology, bioengineering, genetic engineering, computer science, and information technology. It focuses on designing, constructing, and reworking both new and existing biological parts, devices, and systems. The goal of this field is to improve biological functions to tackle pressing issues such as global health, climate change, food insecurity, and energy challenges. Key areas in synthetic biology include:

- **Design of Biological Systems**: Synthetic biology seeks to design and build new biological parts (such as DNA sequences), devices (like genetic circuits), and systems (such as whole organisms). It applies principles from engineering, such as standardization, modularity, and abstraction, to make the design process more predictable and scalable. One real-world example of this is the development of synthetic insulin. Scientists have designed and built DNA sequences that match the human insulin gene, which they insert into bacteria. The bacteria then produce human insulin, which can be harvested and used to treat diabetes.[1]

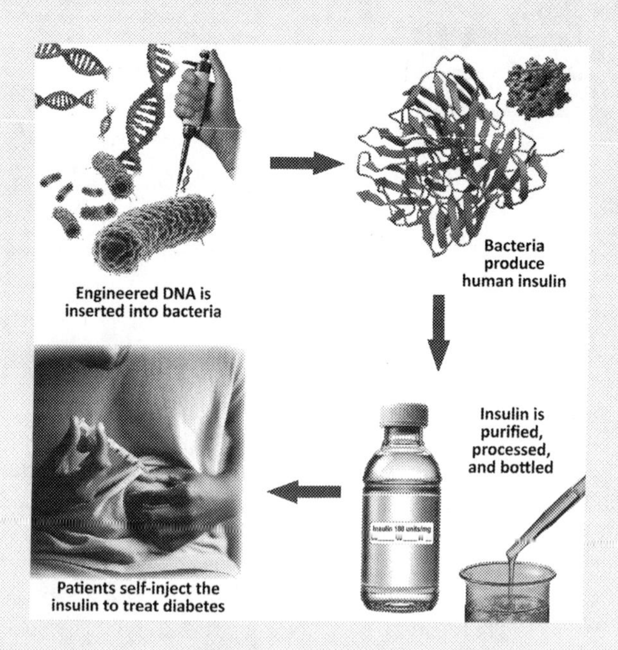

- **Redesign of Existing Biological Systems**: This involves modifying existing biological systems to perform new functions. An example is the development of genetically modified (GM) crops. For instance, scientists have modified corn plants to express a gene from the bacterium *Bacillus thuringiensis* (*Bt*) that produces a toxin lethal to certain pests. This *Bt* corn is more resistant to these pests, reducing the need for chemical pesticides.[2]

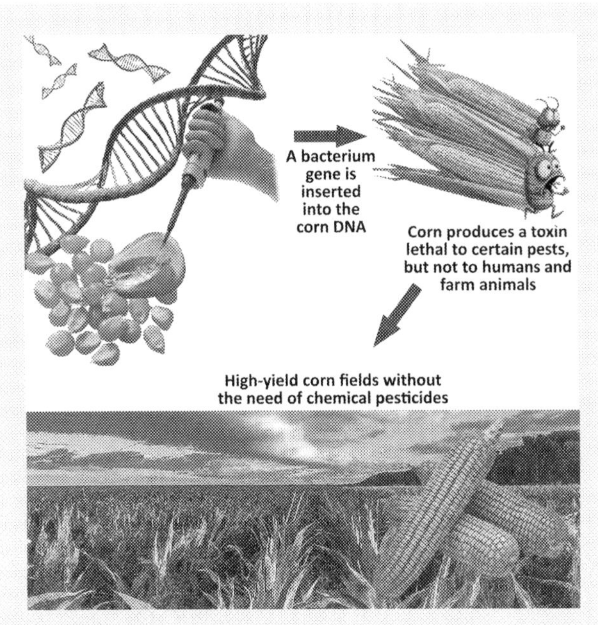

- **Prototyping and Testing**: Like other forms of engineering, synthetic biology involves iterative cycles of design, construction, testing, and analysis. Rapid prototyping and testing techniques are used to refine designs and improve functionality. An example of this is the process of such as a modified yeast strain, for biofuel production. The initial design is constructed and then tested to see if it effectively produces the desired product (like ethanol). If not, the design is refined and tested in iterative cycles until the optimal performance is achieved.[3]

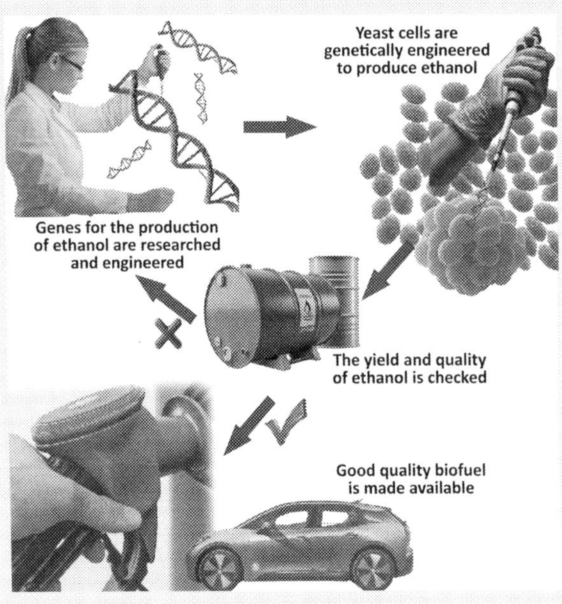

- **Bioinformatics and Computational Modeling**: Synthetic biology leverages the power of computers and software for designing biological systems. This can involve computational modeling of biological systems, machine learning techniques to predict how modifications affect system behavior, and bioinformatics tools for managing and analyzing large amounts of biological data. For example, the Human Genome Project, which involved sequencing and analyzing the entire human genome, heavily relied on bioinformatics and computational modeling. These tools helped scientists predict the function of different genes and their potential interactions based on their sequences and other data.[4]

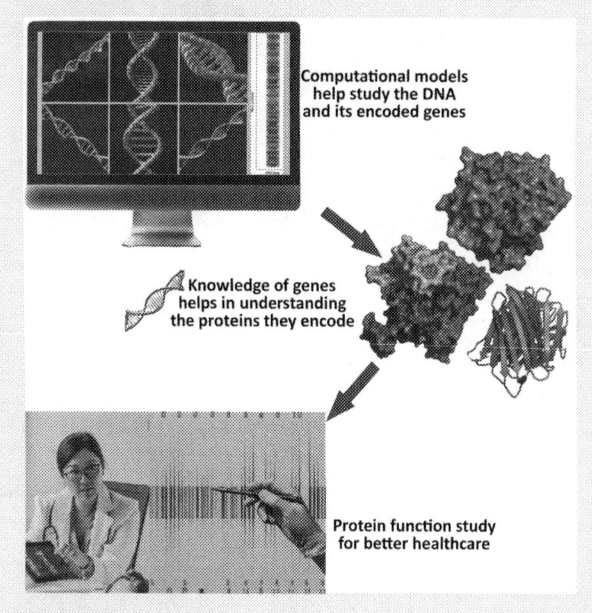

- **Biological Machines and Nanotechnology**: Some researchers in synthetic biology are interested in creating biological machines or devices at the nanoscale. These could have applications in areas such as drug delivery, biomedical devices (including implantable biosensors or nanobots), environmental remediation and sensing, carbon capture, or data storage, to name a few. A practical example is the development of targeted drug delivery systems using nanotechnology. For example, researchers are developing liposomes, tiny "biological machines," that can encapsulate drugs and deliver them directly to cancer cells, thus increasing the efficacy of the treatment and reducing side effects.[5]

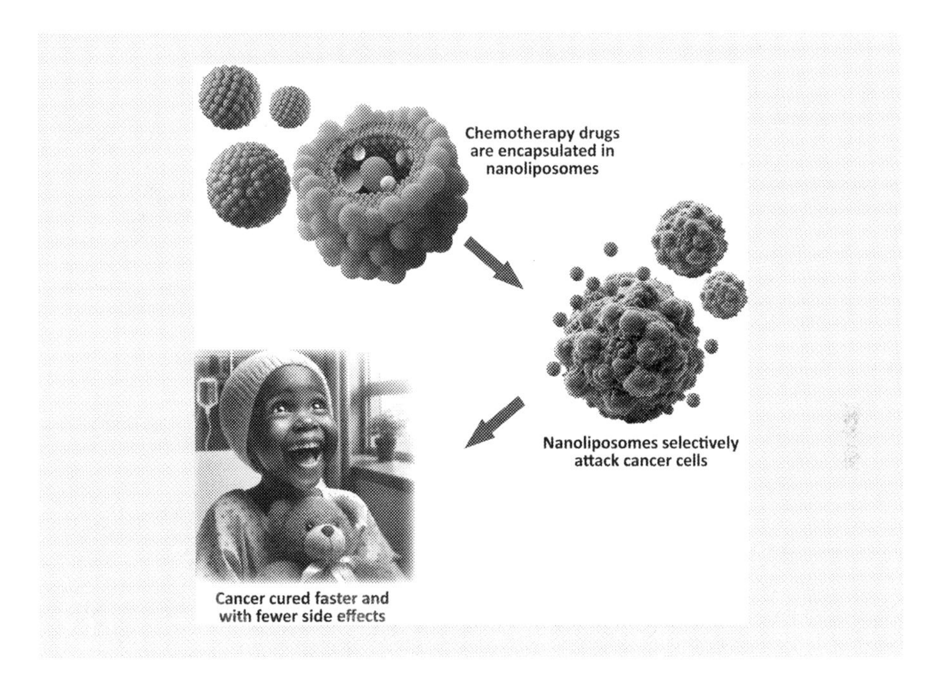

This fusion of innovative research and production techniques has not only sparked a wave of new scientific and medical breakthroughs, but has also delivered practical products that challenge current knowledge boundaries. However, these developments carry with them significant social impacts that must be considered by the wider community.

- **Applications**: Synthetic biology has a wide range of potential applications across many areas and industries. These include producing biofuels and renewable chemicals, developing new medical treatments and diagnostics, creating novel materials, and remediating environmental damage. One example is the development of the mRNA COVID-19 vaccines by Moderna and Pfizer/BioNTech. Synthetic biology techniques were used to construct mRNA sequences that instruct cells to produce a harmless piece of the coronavirus spike protein, triggering an immune response.[6]
- **Ethical, Legal, and Social Implications**: It is crucial that we collectively think about how to ethically and responsibly proceed with expanding and harvesting this technology. Given its potential to profoundly alter biological systems, synthetic biology raises many ethical, legal, and social questions. These include questions about biosafety, biosecurity, intellectual property, regulation, and the potential impacts on society and the environment. A real-world example of this is the debate over CRISPR gene-editing technology. With unprecedented precision, this technique can alter the DNA of organisms, including humans, raising complex questions about biosafety, biosecurity, intellectual property rights, and potential societal and environmental impacts.[7] These topics will be discussed more at length in Section 4.10.

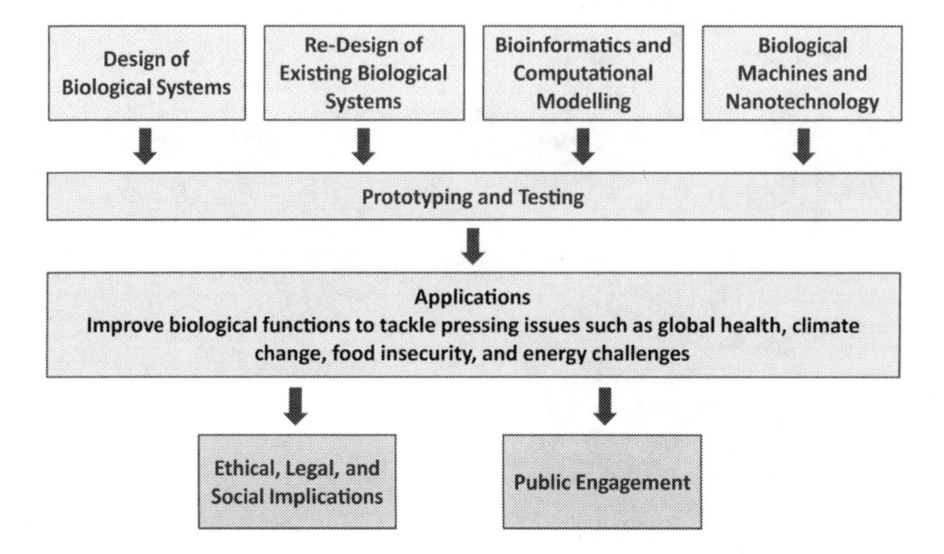

FIGURE 1.1 Overview of synthetic biology and its key research area, application, and consequences.

- **Public Engagement**: Because of the far-reaching implications of synthetic biology, there is a strong need for public dialogue and engagement. This helps ensure that societal values and concerns are considered in the decision-making processes. The UK's Human Fertilization and Embryology Authority (HFEA) provides an example. The HFEA held a public consultation on mitochondrial donation, a new *in vitro* fertilization (IVF) technique that could prevent mitochondrial diseases but involves modifying the human germline. The consultation helped inform the HFEA's decision-making and made the United Kingdom the first to legally approve this technique.[8]

The promise of synthetic biology is vast, offering potential solutions to some of the world's most pressing challenges. However, the field also raises important questions about biosecurity and regulation that must be carefully considered as it develops (Figure 1.1).

1.2 THE EMERGENCE OF BIOSECURITY CONCERNS

Biosecurity concerns emerge in synthetic biology primarily due to the power and accessibility of the technologies involved. While synthetic biology holds great promise for beneficial applications, it also has the potential to be misused, either unintentionally or intentionally, which can pose risks to public health, the environment, and national and global security. The reasons for these concerns include:

- **Creation of Pathogens**: Synthetic biology could be used to create or modify harmful biological agents, such as viruses or bacteria. This could include recreating known pathogens from scratch, enhancing the virulence or transmissibility of existing pathogens, or engineering entirely new organisms with harmful properties. For example, a virology research group synthesized the horsepox virus, a relative of smallpox, using genetic pieces ordered in the mail. While horsepox is not known to harm humans and is not considered a major agricultural threat, the technique used in the study could potentially be applied to recreate smallpox, making it a technique of concern. This highly dangerous disease was declared eradicated in 1980. The study raises concerns about the risks associated with synthetic biology and the potential for the creation or modification of harmful biological agents.[9]
- **Dual-Use Research of Concern (DURC)**: Many synthetic biology research areas could have both beneficial and harmful applications, making them "dual use." For example, a technique developed to engineer viruses for therapeutic purposes could also create harmful viruses. Gain-of-function (GOF) research in virology provides an example of dual use. This research involves modifying viruses to make them more virulent or more transmissible, with the benevolent intention of understanding how transmission occurs or to develop novel vaccines. However, the same knowledge and techniques could potentially be misused to create more harmful viruses with malintent of using them for bioterrorism. A typical everyday example of dual use is provided by a kitchen knife, which is an ordinary household item that is designed for cutting food and preparing meals. However, a knife could also potentially be used as a weapon to harm others. So, while the knife's primary purpose is as a kitchen tool, it has a dual use as both a cooking implement and a potential weapon. Similarly, many products of synthetic biology could result in "dual use." An example of GOF research that falls under DURC is controversial experiments on H5N1 bird flu (discussed in more detail in Section 1.4).[10]
- **Bioterrorism and Biological Warfare**: The same technologies that enable advances in synthetic biology could be exploited by malicious actors for bioterrorism or biological warfare. This is a particular concern given the increased accessibility of these technologies to other scientists who are not experts in synthetic biology. A historical example is the use of aerosolized *Bacillus anthracis* (anthrax) spores in the 2001 anthrax attacks in the United States, where 5 people died and 17 were infected with the modified anthrax strain.[11] The ripple effects of this event are still visible today, as it resulted in the creation of the U.S. Federal Select Agent Program (see Section 1.4 for more details).[12] Today, advances in synthetic biology could potentially allow a bioterrorist to engineer more harmful biological weapons than the 2001 Anthrax incident.[13]

- **Accidental Release**: There is a risk that GMOs created through synthetic biology could accidentally escape from a laboratory setting and have unforeseen impacts on natural ecosystems or human health. An example of accidental release is the case of GM rapeseed.[14] During transportation in several countries, incidents of unintentional environmental release of GM rapeseed have occurred. These releases were due to seed spillage during the transport or import of rapeseed, both in countries cultivating GM rapeseed and those that do not. Although there have been no reports of serious detrimental consequences resulting from these releases, the high potential for hybridization and gene transfer within the genus, particularly with *Brassica rapa*, raises concerns about potential environmental and agricultural perturbations. Therefore, specific programs should be implemented to effectively monitor and manage the accidental release of GM crops to mitigate any unforeseen impacts on ecosystems and agriculture.
- **Unintended Consequences**: Unintended consequences are a significant aspect of the biosecurity concerns arising from synthetic biology. The power and accessibility of the technologies involved in this field increase the likelihood of unintended outcomes. In the pursuit of beneficial applications, there is always a risk of unforeseen repercussions that could affect public health, the environment, and national security. The complex nature of genetic manipulation and the intricate interplay between organisms and ecosystems make it challenging to predict the full extent of the consequences that may arise from synthetic biology. Accidental releases, unintended gene transfers, and the creation of unforeseen pathogens are potential outcomes that demand careful consideration and risk mitigation strategies. It is crucial to exercise caution and implement robust biosecurity measures to minimize the likelihood of unintended consequences and ensure the responsible development and application of synthetic biology technologies. For example, the intentional release of GM mosquitoes in Brazil by a biotech company is an example of unintended consequences. These mosquitoes were designed to control the population of *Aedes aegypti* mosquitoes, which spread diseases like Dengue and Zika. However, after release, it was found that the GM mosquitoes were able to breed in the wild, contrary to expectations.[15]
- **Intentional Release**: The deliberate release of synthetic organisms or materials that are biohazardous is a significant issue in biosecurity. This includes the potential abuse of synthetic biology. People may create and intentionally release harmful biological agents. This could happen for several reasons. Some might have malicious intent. Others may have good intentions but make misguided efforts to address a problem. Such agents could include deadly pathogens or organisms designed to disrupt ecosystems or agricultural systems. These actions could lead to public health crises, environmental damage, or economic loss. Furthermore, as synthetic biology becomes more accessible, the risk of bioterrorism increases. Safeguarding against the intentional release of pathogens or other synthetic biology products requires stringent regulation, robust surveillance, effective law enforcement, and international cooperation

to share knowledge and best practices. A hypothetical scenario could involve a malicious actor intentionally releasing a genetically engineered pathogen to cause harm. This might be a bacteria designed to devastate crops, for instance, causing widespread famine and economic damage.

- **Do-It-Yourself (DIY) Biology**: With the rise of DIY biology communities and home biology kits, there is a potential for misuse of synthetic biology techniques outside of traditional laboratory settings, where there is less oversight and control by biosecurity experts and committees. An example is the case of a biohacker who injected himself with a homemade herpes treatment in front of a live audience in 2018. This event raised concerns about safety and the lack of oversight in DIY biology communities.[16]
- **Lack of Regulation**: As a new and rapidly advancing field, synthetic biology often falls outside the scope of existing regulations, raising concerns over oversight and control. For example, until very recently, the use of gene-editing tools such as CRISPR in human embryos was an unregulated area in many countries, raising significant concerns about the potential for uncontrolled genetic modifications.
- **Ethical, Social, and Economic Implications**: The landscape of synthetic biology extends far beyond physical safety and security, engendering many ethical, social, and economic implications. One such implication lies in the potential use of this technology to alter human genetics, a prospect that conjures a labyrinth of complex ethical conundrums. The growing disparity in global access to these sophisticated technologies is a rapidly emerging concern. Countries with substantial resources can readily tap into the advantages of synthetic biology, while nations with limited resources may grapple with restricted or no access. This discrepancy is clearly seen in access to life-saving vaccines developed using synthetic biology techniques. For example, mRNA vaccines for COVID-19, which were produced rapidly and proved highly effective, were initially inaccessible to low-income countries due to their high cost and stringent storage requirements. This exacerbated the global health disparity during the COVID-19 pandemic (2020–2023).[17]

  Similarly, revolutionary cancer treatments, such as CAR-T cell therapy, are also out of reach for many patients in resource-limited settings.[18] This therapy, which involves engineering a patient's immune cells to fight cancer, is expensive and requires sophisticated medical infrastructure, making it available only in wealthy countries.

  This unequal access to the benefits of synthetic biology can compound existing global inequities and forge new forms of social and economic divides. Therefore, this issue is not just a technical one, but a profound ethical and social concern that warrants immediate attention and policy intervention. Developing and implementing policies to address these disparities and ensure equitable access to the benefits of synthetic biology is a crucial challenge that the global community needs to tackle urgently.

As synthetic biology continues to advance, appropriate measures must be put in place to manage and mitigate biosecurity risks. This includes technical measures, such as developing containment and deactivation strategies for synthetic organisms, and policy measures, like establishing clear regulations and oversight mechanisms.

1.3 NOTABLE CASE STUDIES IN BIOMEDICAL SYNTHETIC BIOLOGY

Synthetic biology, when applied to biomedicine, has a tremendous potential to revolutionize different subfields of medicine, provide new treatments for hard-to-treat diseases and infections, and dramatically improve and extend life. There are numerous case studies that highlight the field's potential benefits and the importance of applying rigorous biosafety and biosecurity practices. A few notable examples include:

- **Synthetic Artemisinin**: In the early 2000s, a team of scientists engineered yeast to produce artemisinin, a drug used to treat malaria. This work demonstrated the potential of synthetic biology to produce valuable drugs more sustainably and cost-effectively than traditional methods. However, the project also faced challenges, including regulatory hurdles and resistance from farmers who traditionally grew the plant that naturally produces artemisinin.[19]
- **The Creation of Synthetic Genomes**: In 2010, a team created the first synthetic cell by constructing a copy of a bacterial genome and inserting it into another bacterium. This landmark achievement raised many safety and ethical concerns, including the potential risks of creating new life forms and the potential for misuse of the technology.[20]
- **The Genetic Engineering of Immune Cells**: CAR-T cell therapy, which involves genetically engineering a patient's own T cells to fight cancer, has shown remarkable success in treating certain types of cancer, particularly childhood leukemias that previously would have been terminal. Thus, saving lives of children that would have been otherwise condemned to death. However, the process also poses risks, such as severe immune reactions and the potential for the modified cells to become cancerous.[21]
- **The Development of Gene Drives**: Gene drive experiments use a technique for spreading specific genetic traits through populations more rapidly than would occur naturally. While they hold promise for addressing issues like mosquito-borne diseases, they also pose significant ecological risks if they spread uncontrollably.[22]
- **The CRISPR-Babies Controversy**: In 2018, a Chinese researcher announced the birth of twin girls whose genomes had been edited using CRISPR-Cas9 technology. The experiment, which was widely condemned by the scientific community and caused international public outrage, highlighted the potential for misuse of powerful genetic technologies and underscored the need for strong ethical guidelines and oversight in synthetic

biology.[23] Since then, the lead scientist has been discredited and outcast by the scientific community. However, the long-term effects, if any, of the experiment on the children remain to be seen.

Each of these cases illustrates the promise that synthetic biology holds for improving the future of humanity through medicine, as well as the importance of having robust biosafety and biosecurity measures in place. They illustrate the importance of careful consideration of ethical implications and clear communication with the public prior to developing and implementing synthetic biology technologies in the biomedical field.

1.4 HISTORICAL BIOSECURITY INCIDENTS: LESSONS AND CONSEQUENCES

Laboratory biosecurity is part of biorisk management, and it focuses on protecting the biological agents from intentional or unintentional theft, loss, or misuse (biosecurity will be covered in depth in Chapters 8–11). The history of biosecurity is filled with lessons that can inform safe practices in synthetic biology. Here are a select few examples that highlight different vulnerabilities that are still relevant today:

- **The 1971 Aralsk Smallpox Outbreak**: In 1971, Aralsk, a Kazakh town in the former Soviet Union, experienced a smallpox outbreak. This accident resulted from the virus being tested as a potential biological weapon (BW) in the former Soviet Union's BW program. Ten people were infected, and three died from hemorrhagic smallpox. Authorities contained the outbreak through quarantine and vaccination. However, the former Soviet Union successfully hid this outbreak. They also concealed decades of work on weaponizing smallpox, as well as other outbreaks in the 1970s and 1980s. This concealment reveals a systemic weakness in the biological disarmament regime. The incident showed that the Biological Weapons Convention (BWC, see Section 5.1.1) was – and still is – not effective in ensuring countries comply with its rules. After the Soviet BW program was exposed, the international community did not impose sanctions. Until now, no individual, group, or country involved in the development, production, or stockpiling of biological weapons has faced international reprimand.[24] This event also highlights the risks of working with dangerous pathogens, even in controlled environments. It has led to the implementation of stricter controls on such research worldwide.[25]
- **The 1979 Sverdlovsk Anthrax Leak**: Similarly, an accidental release of anthrax spores from a Soviet military research facility caused at least 66 deaths in 1979. Although the Soviet Union initially claimed that the outbreak was due to contaminated meat, it was later determined to be the result of the BW program testing. It took years to uncover the extent of the activities that led to this outbreak,[26] highlighting the need for stronger regulations that foment transparency in dealing with biosecurity incidents globally.[27]

- **The 2001 Anthrax Attacks**: In the wake of the September 11, 2001 attack in the United States, letters laced with anthrax spores were mailed to news media offices and 2 U.S. Senators, causing 5 deaths and 17 people to fall ill with a modified *Bacillus anthracis* (anthrax) strain.[11] The FBI traced the source of the anthrax to the U.S. Army Laboratory USAMRIID, using a unique genetic signature contained in the spores. This allowed investigators to pinpoint the source of the spores to a single flask within the army lab that was under the control of a single person, the main suspect in this case. It highlights the importance of insider threats and having a robust personnel reliability program. The ripple effects are still visible today. As a result, the U.S. Government created the Federal Select Agents and Toxins Program,[12] and heightened biosecurity measures were implemented in labs working with such select agents.[13]
- **The 2007 Foot and Mouth Disease (FMD) Leaks in the United Kingdom**: In 2007, the United Kingdom faced a serious biosecurity lapse when the Foot and Mouth Disease (FMD) virus leaked from a vaccine production facility and a connected government research lab. The cause was deteriorated pipes that should have safely transported the virus to a treatment tank located miles away. Instead, the neglected pipes allowed the virus to seep into the surrounding ground. Heavy rainfall likely exacerbated the situation, spreading the contamination to nearby farms through the soil.[28] Subsequent investigations identified weaknesses in the biosecurity protocols, including the lack of maintenance of the old infrastructure and a lack of proper oversight. This incident prompted the United Kingdom to revise its biosecurity regulations and practices to ensure that they were in alignment with European Union standards.[29]
- **The H5N1 and H7N9 Avian Influenza Research Controversy (2011–2014)**: From 2011 to 2014, research aimed at making the H5N1 and H7N9 avian influenza strains more transmissible in mammals stirred international controversy. This GOF research, which can be used for both beneficial and potentially harmful purposes (dual-use research of concern or DURC), was initially greenlit by U.S. authorities. However, widespread public concern led to the halt of such research. This pause prompted the United States to refine its oversight of dual-use research.[30,31] By 2017, after thorough review by a dedicated committee, the National Institutes of Health (NIH) resumed funding for this line of research, starting with a single, rigorously evaluated grant.[31]
- **The COVID-19 Pandemic (2020–2023)**: The COVID-19 Pandemic has been covered at length in other academic publications and thus will not be discussed here in detail. While the origins of the SARS-CoV-2 virus remain under investigation, the pandemic has underscored the devastating impact of a novel pathogen and highlighted the importance of robust public health, scientific collaboration, and transparency in addressing such a global crisis.[32]

These incidents underscore the importance of implementing and maintaining a robust biorisk management (BRM) program, which will be discussed at length in

Chapter 7, including strong biosafety and biosecurity measures, when working with biological materials, whether naturally occurring or synthetically created. They also highlight the need for transparency, collaboration, and public trust in managing biosecurity risks.

1.5 RECENT ACCOMPLISHMENTS IN SYNTHETIC BIOLOGY AND BIORISK GOVERNANCE UPDATES

The following timeline, from 2001 to 2023, details key milestones in synthetic biology, illustrating the field's rapid evolution and its significant impact on the frontier of scientific knowledge. Each milestone underlines the growing importance of careful oversight and regulation in the field. This chronological overview also includes major achievements in biorisk management. Together, these milestones trace the development of both synthetic biology and biorisk management, emphasizing their combined importance for the responsible advancement of the field, as summarized in Figure 1.2 (2001-14) and Figure 1.3 (2015-23).

- **2001: First Synthetic Poliovirus Created –** Researchers demonstrate the ability to synthesize a complex viral genome, such as the poliovirus, in its entirety, from scratch by following the virus known genetic sequence. This was a pivotal moment in synthetic biology, showing that viruses could be created without natural templates, raising both possibilities for new therapeutic developments and biosecurity concerns.[33]

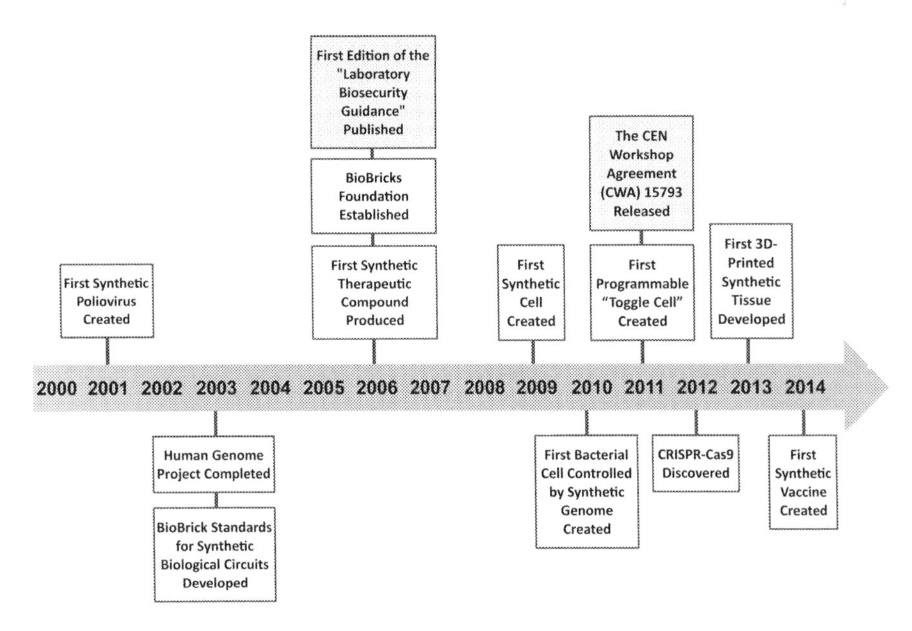

FIGURE 1.2 A timeline of synthetic biology and biorisk governance accomplishments from 2000 to 2014.

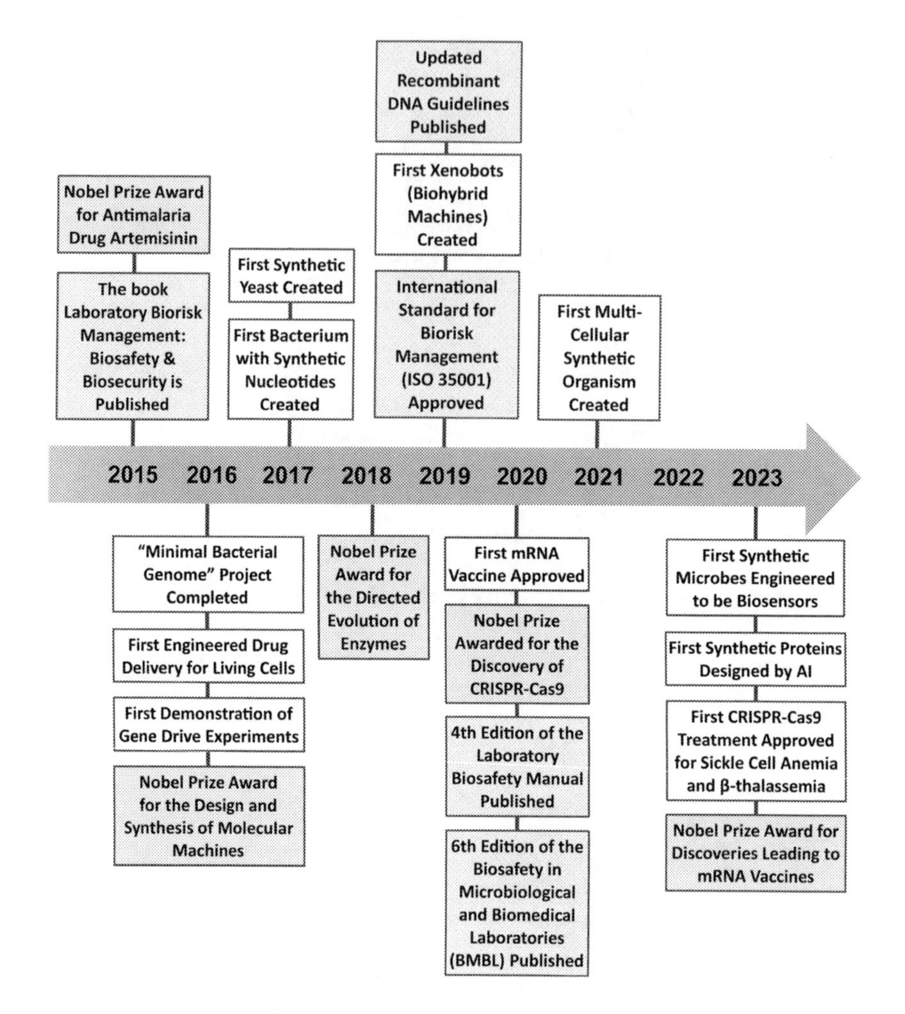

FIGURE 1.3 A timeline of synthetic biology and biorisk governance accomplishments from 2015 to 2023.

- **2003: Human Genome Project Completed** – The completion of the Human Genome Project was a seminal event in the biosciences, providing the first complete human genome sequence.[4] This landmark achievement laid the groundwork for synthetic biology to experiment with precise human DNA modifications, paving the way for new techniques to design and build custom DNA circuits.
- **2003: BioBrick Standards for Synthetic Biological Circuits Developed** – The creation of the BioBrick standard resulted in a unified framework for constructing, assembling, and sharing DNA sequences, thus streamlining the creation of synthetic biological circuits.[34]

- **2006: First Synthetic Therapeutic Compound Produced** – Researchers engineered a yeast strain to produce artemisinin, a drug for treating malaria. This was the first example of using synthetic biology to produce a valuable therapeutic compound, a leap forward in drug manufacturing.[35]
- **2006: BioBricks Foundation Established** – The BioBricks Foundation was established to support and promote the standardization of biological parts, a key element for the modular and systematic design of synthetic biology in a responsible way.[36]
- **2006: First Edition of the WHO "Laboratory Biosecurity Guidance" Published** – This document provides guidance on the implementation of biosecurity measures in laboratories to prevent the loss, theft, misuse, or unauthorized access to biological materials that could be potentially harmful. This guidance marked a significant step forward in providing recommendations for enhancing security measures in traditional biology laboratory settings. See Section 7.1 for further information.
- **2009: First Synthetic Cell Created** – A bacterial cell was created using a chemically synthesized genome, showcasing the ability to construct a living organism with a chemically synthesized genome, which challenges our understanding of life's definition by blurring the lines between natural and synthetic life forms.[20]
- **2010: First Bacterial Cell Controlled by Synthetic Genome Created** – For the first time, a synthetic genome was transplanted into a recipient cell, resulting in a fully functional organism.[20] This achievement demonstrated that genomes designed and constructed in the lab could be booted up in a cellular environment, essentially creating new life forms from synthesized DNA. This milestone was critical as it showed that life could be not just edited but fundamentally created through synthetic biology.
- **2011: First Programmable "Toggle Cell" Created** – A genetic "toggle switch" was shown to enable cells to toggle between different states of gene expression, a fundamental step for creating programmable cells.[37] This refinement of synthetic genetic circuits allows for more complex behaviors to be encoded into synthetic organisms.
- **2011: The CEN Workshop Agreement (CWA) 15793 Released** – The CWA 15793 is a reference document representing a set of voluntary guidelines central to biorisk management within laboratory settings. It details a comprehensive set of requirements for the handling and storing of biological agents and toxins, thereby playing an instrumental role in biosafety and biosecurity. See Section 7.2 for further information.
- **2012: CRISPR-Cas9 Discovered** – The CRISPR-Cas9's discovery provided a transformative genome editing tool, allowing precise and efficient genetic alterations, catalyzing advancements across all life sciences and medicine. This powerful tool has revolutionized synthetic biology, making it possible to precisely edit DNA with unprecedented ease.[38]
- **2013: First 3D-Printed Synthetic Tissue Developed** – The construction of synthetic tissue structures using 3D printing techniques marked a confluence of synthetic biology with tissue engineering, paving

the way for creating functional biological tissues using 3D printing technologies.[39]

- **2014: First Synthetic Vaccine Created** – Researchers designed and created synthetic peptides that could stimulate the immune system, offering potential for rapid and precise responses to emerging infectious diseases [40]
- **2015: Nobel Prize Award for Antimalaria Drug Artemisinin** – Dr. Tu Youyou shared half of the Nobel Prize in Physiology or Medicine. Her research led to the development of the antimalaria drug artemisinin using synthetic biology techniques.[41] The other half of the Nobel Prize was awarded jointly to William C. Campbell and Satoshi Ōmura for their discoveries concerning a novel therapy against infections caused by roundworm parasites.
- **2015: The book *Laboratory Biorisk Management: Biosafety & Biosecurity* is Published** – This is the first publication proposing a biorisk management approach.[42]
- **2016: "Minimal Bacterial Genome" Project Completed** – The "Minimal Genome Project" created the first living cell with a genome of only 531,000 base pairs and just 473 genes, making it the smallest genome of any self-replicating organism.[43] The project successfully identified the smallest number of genes necessary for life. Insights from this project have significant implications for constructing synthetic cells and understanding biological complexity.
- **2016: First Engineered Drug Delivery for Living Cells** – Synthetic biology researchers develop new methods for engineering living cells to deliver drugs to specific organs and tissues in the body. This technology has the potential to revolutionize the way we treat cancer and other diseases.[44]
- **2016: First Demonstration of Gene Drive Experiments** – The first concept of gene drives was demonstrated, allowing engineered genes to propagate specific genetic traits through insect populations, a powerful technology with profound ecological and biosecurity implications.[45]
- **2016: Nobel Prize Award for the Design and Synthesis of Molecular Machines** – Three scientists, Jean-Pierre Sauvage, Fraser Stoddart, and Bernard Feringa, shared the Nobel Prize in Chemistry. Their research led to the design and development of molecular machines, which is a topic closely related to synthetic biology.[46]
- **2017: First Bacterium with Synthetic Nucleotides Created** – A semi-synthetic bacterium strain was created with a six-letter genetic code (instead of the natural four letters, A, T, G, C) by adding two synthetic bases, X and Y. This expanded "genetic alphabet" could potentially lead to the production of novel proteins and biomaterials.[47]
- **2017–2018: First Synthetic Yeast Created** – Development of synthetic genomes led to the creation of a synthetic version of the yeast *Saccharomyces cerevisiae*'s genome. This effort, known as the Synthetic Yeast 2.0 Project, aims to completely construct a designer eukaryotic genome from scratch, which can lead to a better understanding of genome

structure and function, and could even result in yeast strains with desirable industrial properties.[48]

- **2018: Nobel Prize Award for the Directed Evolution of Enzymes** – Professor Frances H. Arnold was awarded the Nobel Prize in Chemistry. Her research led to the discovery of "directed evolution of enzymes," a technique that is widely used in synthetic biology for the creation and modification of biological parts.[49]
- **2019: International Standard for Biorisk Management (ISO 35001) Approved** – This ISO standard provides a systematic approach to manage biological risks associated with laboratories and other related organizations. It is particularly relevant in ensuring biosafety and security in environments dealing with biological agents, thus contributing significantly to the field of biosecurity, especially in the context of synthetic biology. See Section 7.3 for further information.
- **2019: First Xenobots (Biohybrid Machines) Created** – The first living, programmable organisms called "xenobots" were created, which are tiny biohybrid machines that can move, push payloads, and even exhibit collective behavior in the presence of a swarm.[50] This development opened up new possibilities for living machines that could perform tasks such as targeted drug delivery.[51]
- **2019: Updated Recombinant DNA Guidelines Published** – The most recent version of the NIH Guidelines for Research Involving Recombinant or Synthetic Nucleic Acid Molecules aims at ensuring the safe and responsible conduct of research involving recombinant DNA technology. See Section 5.2.5 for further information.
- **2020: First mRNA Vaccine Approved** – The first mRNA vaccines were approved to provide immunity against SARS-CoV-2, showing how powerful synthetic biology can be in dealing with major health crises. Unlike traditional vaccines, these use a piece of genetic material, mRNA, to make our bodies build a protein that triggers an immune response. This approach was a significant leap forward because it was developed rapidly in less than a year, which is remarkable compared to the usual timeline for vaccine development. This achievement demonstrated the ability to respond to health emergencies quickly and opened up new possibilities for using mRNA tech in other diseases, changing the game in vaccine science.[6]
- **2020: Sixth Edition of the *Biosafety in Microbiological and Biomedical Laboratories* (BMBL) Published** – The sixth edition of the BMBL provides updated guidelines and recommendations for the safe handling of biological materials and is a valuable resource in the field of biosafety and laboratory research. See Section 5.2.3 for further information.
- **2020: Nobel Prize Awarded for the Discovery of CRISPR-Cas9** – Two scientists, Jennifer A. Doudna and Emmanuelle Charpentier, shared the Nobel Prize in Chemistry. Their research led to the development of the CRISPR-Cas9 gene-editing technology, which has become a fundamental tool in synthetic biology for manipulating genetic material.[52]

- **2020: Fourth Edition of the WHO *Laboratory Biosafety Manual* Published –** This updated edition of the manual includes a core document along with seven subject-specific monographs, adopting a risk-based approach to promote optimal resource use and sustainable biosafety and biosecurity practices tailored to the unique situation of each laboratory. The manual is widely used across various levels of clinical and public health laboratories and other biomedical sectors globally. See Section 5.1.4 for further information.
- **2021: First Multicellular Synthetic Organism Created –** Scientists engineered the first multicellular synthetic organism, showing the ability to design and construct complex life forms.[53]
- **2023: First Synthetic Microbes Engineered to be Biosensors –** Synthetic biology researchers develop new methods for engineering microbes to detect and respond to environmental toxins. These advances have the potential to be used to create new types of biosensors and bioremediation systems.[54]
- **2023: First Synthetic Proteins Designed by AI –** The use of artificial intelligence to design synthetic proteins demonstrated the effectiveness of merging computational modeling and synthetic biology to reach new frontiers in the development of novel therapeutics and diagnostics.[55]
- **2023: First CRISPR-Cas9 Treatment Approved for Sickle Cell Anemia and β-Thalassemia –** This therapy, branded as Casgevy, will be used to treat sickle cell disease and β-thalassemia, two blood diseases that so far had no cure. The approval of this therapy is a significant milestone that will open up the doors for further treatments based on CRISPR technologies.[56]
- **2023: Nobel Prize Award for Discoveries Leading to mRNA Vaccines –** Two scientists, Katalin Karikó and Drew Weissman, shared the Nobel Prize in Physiology or Medicine. Their research led to the discovery of "nucleoside base modifications" that enabled the development of the mRNA vaccines against COVID-19. This work is based on fundamental principles of synthetic biology, such as the design and construction of artificial genetic circuits.[57]

Since the beginning of the 21st century, synthetic biology has experienced exponential growth, with no signs of slowing down its rapid expansion. In tandem, biosafety has generally kept pace, though biosecurity still faces challenges due to a lack of clearly defined, enforceable measures. To address this shortcoming, there is an urgent need for the biorisk management community to update its existing guidelines and proactively create new documents to include up-to-date biosecurity measures. These updates are essential to effectively manage current challenges posed by the latest advancements in synthetic biology. New, forward-looking biorisk guidelines should also be developed to anticipate and keep pace with future challenges in line with the trajectory of synthetic biology research.

1.6 SCOPE AND OBJECTIVES OF THIS BOOK

This book serves as an extensive guide on biosecurity in synthetic biology. It is designed for researchers, biorisk management professionals, policymakers, educators, and the general public who may be interested in this subject. This book further aims to assist the reader with due diligence when conducting biosecurity risk assessments related to laboratory research in the field of synthetic biology. The necessity to present "Biosecurity in the Age of Synthetic Biology" in this book is underscored by several key factors:

- **Education and Awareness**: A significant audience comprising policymakers, educators, and scientists who need more and better understanding of the biosecurity implications inherent in synthetic biology. This book aims to educate a diverse readership about these critical issues.
- **Fostering Responsible Research**: This book encourages a culture of responsibility among synthetic biologists by providing guidelines for safe and ethical research practices, discussing the ethical and societal implications of various topics and case studies, and urging researchers to consider these factors in their work.
- **Policy Guidance**: In the complex landscape of synthetic biology and biosecurity, policymakers require clear and accessible information for informed decision-making. This book is designed to offer valuable guidance, including recommendations for regulatory measures, oversight mechanisms, and other policy interventions.
- **Public Engagement**: Engaging public understanding is pivotal to the responsible evolution of synthetic biology. This book aims to spark public dialogue on biosecurity, guiding discussions on societal values, potential risks, and benefits.
- **Addressing Dual-Use Concerns**: The dual-use nature of synthetic biology, wherein research can serve both beneficial and harmful purposes, presents complex issues. This book delves into these challenges and proposes methods to manage dual-use risks.
- **Proactive Approach**: With the rapid progression of synthetic biology, this book's focus on biosecurity takes a proactive approach, aiming to foresee potential risks and develop mitigation strategies before they emerge as pressing problems.
- **Cross-disciplinary Approach**: By its very nature, synthetic biology merges aspects of biology, engineering, computer science, ethics, policy, and more. Reflecting this interdisciplinarity, this book stimulates cross-disciplinary dialogue and collaboration.
- **Global Collaboration**: Biosecurity is a global issue demanding international cooperation. This book aspires to cultivate such cooperation, offering a platform to share perspectives, best practices, and lessons learned from around the globe.

1.7 CONCLUDING THOUGHTS

Synthetic biology is a field rife with groundbreaking scientific progress, and along with it come significant biosecurity challenges and responsibilities. This field's ability to redesign and construct biological systems offers transformative solutions for pressing global issues like health crises, environmental degradation, and energy scarcity. However, the same tools that promise a brighter future also pose risks, such as bioterrorism, accidental release of GMOs, or unintended ecological consequences. This duality emphasizes the need for robust biosecurity measures within the framework of synthetic biology. A collaborative, interdisciplinary approach involving scientists, policymakers, and biosecurity experts is essential. By embracing collaboration, we can leverage synthetic biology's immense potential while mitigating its inherent risks, thus guiding this powerful new technology toward societal betterment and environmental protection. Proactive, collaborative strategies are vital for mitigating biosecurity risks, with various stakeholders responsible for steering this promising field toward beneficial and safe outcomes.

1.8 KEY TAKEAWAYS

- **Interdisciplinary Nature**: Synthetic biology merges diverse fields like biology, bioengineering, and computer science, leading to innovative solutions for health, environment, and energy challenges.
- **Biosecurity Concerns**: These include the potential for misuse of synthetic biology in bioterrorism, accidental release of harmful agents, and unintended ecological impacts.
- **Need for Updated Regulation and Oversight**: The rapid advancement of synthetic biology outpaces existing regulations, calling for updated and comprehensive biosecurity measures.
- **Dual-Use Dilemma**: Synthetic biology's applications have both beneficial and potentially harmful uses, necessitating careful consideration and management of dual-use research.
- **Ethical and Social Considerations**: The field raises complex ethical questions, especially concerning human genetics and equity in access to synthetic biology innovations.
- **Collaborative Approach**: Addressing biosecurity in synthetic biology requires collaboration among scientists, policymakers, ethicists, and security experts.
- **Timeline of Progress in Synthetic Biology**: Real-world examples illustrate both the promises and challenges of synthetic biology, from medical breakthroughs to biosecurity incidents.

1.9 THOUGHT-PROVOKING QUESTIONS

1.1 What are the potential impacts of synthetic biology on global health, climate change, and energy challenges?

1.2 How can the design and redesign of biological systems in synthetic biology be regulated to ensure biosecurity?

1.3 What role does bioinformatics and computational modeling play in synthetic biology, and what are the biosecurity considerations in their use?

1.4 How can the development of biological machines and nanotechnology in synthetic biology be managed to prevent potential misuse or unintended consequences?
1.5 What are the potential applications of synthetic biology, and what are the ethical, legal, and social implications of these applications?
1.6 How can public perception and understanding of synthetic biology be improved to ensure the responsible development and use of this technology?
1.7 How could the current biosafety and biosecurity policies and guidelines have prevented historical incidents like the 2001 Anthrax incident?
1.8 What are the most critical lessons from past biosecurity incidents that still need to be addressed in the current biosafety and biosecurity guidelines?

2 Synthetic Biology and Its Role in Biomedicine

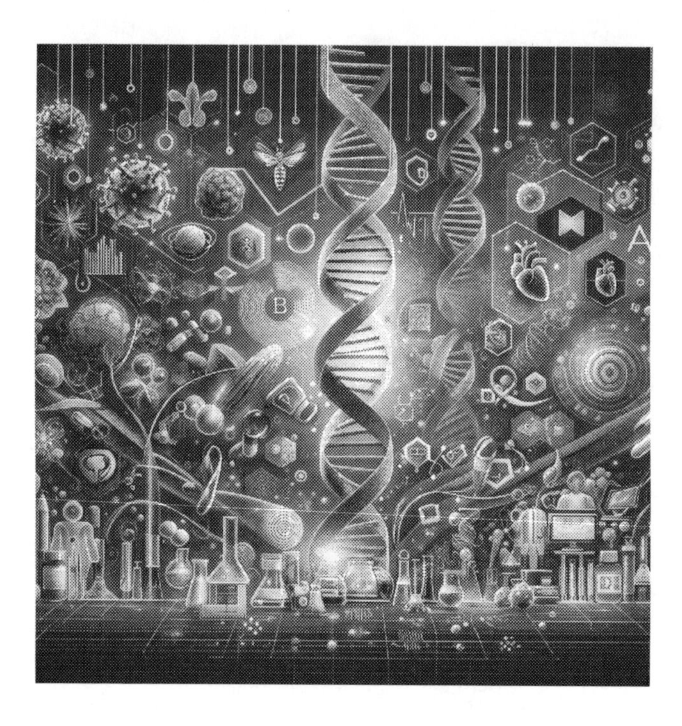

Synthetic biology has been pivotal in advancing biomedicine, illustrating how this innovative field revolutionizes healthcare and medical research. This chapter explains the various applications of synthetic biology in biomedicine, highlighting its potential to create breakthrough treatments and diagnostic tools and offering insights into the future of medical science. This chapter defines synthetic biology, outlining its principles of redesigning and constructing biological systems for specific purposes. It then transitions to the applications of synthetic biology in biomedicine, highlighting key areas such as drug discovery and development, personalized medicine, tissue engineering, and the production of novel therapeutic agents like biosynthetic drugs and vaccines. It addresses concerns such as biosafety, biosecurity, and the ethical implications of genetic manipulation. The need for robust regulatory frameworks to ensure the safe and responsible development of biomedical applications is emphasized. The chapter also looks at the future directions of synthetic biology in biomedicine, speculating on its potential to further revolutionize healthcare. It highlights synthetic biology as a pivotal and rapidly evolving field in biomedicine, with vast

DOI: 10.1201/9781003423171-2

potential to impact healthcare and medical research. It calls for continued exploration and responsible development of synthetic biology applications, ensuring that they are leveraged for the maximum benefit of human health and well-being.

2.1 DEFINITIONS AND KEY CONCEPTS

Synthetic biology is an interdisciplinary field that involves the application of engineering principles to biology. It aims to design and construct new biological parts, devices, and systems or redesign existing natural biological systems for useful purposes. Some key definitions and concepts in the field include:

- **BioBrick**: A BioBrick is a DNA fragment that functions as a biological module. BioBricks can be combined to build larger DNA sequences that perform complex functions. For example, the iGEM (International Genetically Engineered Machine) competition encourages university students to create BioBricks for new functions, such as designing bacteria that can change color in response to different environmental pollutants.[58]
- **Biodesign**: "Biodesign" is the process of designing, testing, and iterating genetic sequences to achieve a desired function or product. This involves using software tools to design new sequences, and lab techniques for their testing. For example, Ginkgo Bioworks is a company that uses biodesign to engineer microorganisms to produce various products, such as flavors, fragrances, and probiotics.[59]
- **Chassis organism**: A chassis organism is a microbial cell deliberately selected and modified to accommodate synthetic systems. *Escherichia coli* is often used as a chassis organism because it is well-understood and easy to manipulate. For example, it has been engineered to produce human insulin. This bacterium serves as the chassis organism where the human insulin gene is inserted, and the bacterium's cellular machinery is used to produce insulin.[1]
- **Genome Editing**: Genome editing (gene editing) is a group of technologies that allow scientists to change an organism's DNA. These technologies allow genetic material to be added, removed, or altered at locations in the genome. CRISPR-Cas9 is currently the most used system for genome editing. The development of CRISPR babies in China (albeit ethically controversial and largely condemned) provides a real-world example of genome editing, where the genes of human embryos were edited to confer HIV resistance.[23]
- **Genetic Engineering**: This is the direct manipulation of an organism's genes using biotechnology. It is a set of techniques, methods, and technologies that alter an organism's genetic material (DNA) by either removing, altering, or adding DNA sequences. For example, Golden Rice is a variety of rice genetically engineered to biosynthesize beta-carotene, a precursor of vitamin A, in the edible parts of rice. This was done to combat vitamin A deficiency in regions where rice is a staple food.[60]

- **Metabolic Engineering**: Metabolic engineering is optimizing genetic and regulatory processes within cells to increase the cells' production of a specific substance. These techniques are used in synthetic biology to produce substances such as biofuels or drugs. An example is the production of artemisinin, a potent antimalarial drug, by engineered yeast. The yeast was genetically engineered to produce a precursor of artemisinin, increasing the availability and lowering the cost of this important drug.[19]
- **Minimal Genome**: A minimal genome is the smallest set of genes that can support life under lab conditions. Defining a minimal genome is a goal of synthetic biology because it provides a simplified platform for further engineering and understanding. *Mycoplasma laboratorium* is an example of a minimal genome, with only the genes necessary for life present in its synthetic genome.[20]
- **Protocell**: Protocells are self-organizing, self-replicating, and self-producing systems that represent the simplest cell models and can be created in the lab. They are used in synthetic biology as they provide a stepping stone toward creating synthetic life. For example, scientists at Harvard University created synthetic protocells in 2013 that mimic some of the behaviors of living cells, including division-like behaviors and energy production.[61]
- **Synthetic Genomics**: Synthetic genomics is a field of synthetic biology that uses the techniques of genetic engineering to synthesize a complete genome from scratch, insert it into a cell, and create a new, living organism. For example, the first synthetic organism, *Mycoplasma laboratorium*, was created by synthesizing a minimal bacterial genome and introducing it into a bacterial cell, replacing its native DNA.[20]
- **Xenobiology**: Xenobiology is a subfield of synthetic biology that focuses on creating novel organisms and biological components that do not exist naturally. This includes the creation of XNA (xeno nucleic acid), an artificial genetic material not found in nature. The development of "alien" life forms with expanded genetic codes by scientists at Scripps Research Institute, where they incorporated two new, artificial nucleotides into the DNA of a bacterial cell, is an example.[47]

2.2 TOOLS AND TECHNIQUES IN SYNTHETIC BIOLOGY

Synthetic biology incorporates numerous tools and techniques from various disciplines like biology, engineering, computer science, and bioinformatics. Below are some key tools and techniques:

- **Bioinformatics Tools**: Various software and computational tools are used to design genetic sequences, model biological systems, analyze data, and more. This includes tools for pathway analysis, sequence alignment, genome annotation, etc. For example, the online resource BLAST (Basic Local Alignment Search Tool) is used to find regions of local similarity between sequences, comparing nucleotide or protein sequences to sequence databases and calculating the statistical significance.

- **Cloning and Expression Systems**: These are used to insert DNA into cells and get the cells to express the protein the DNA codes for. This can involve using plasmids, viral vectors, and various types of host cells (like bacteria or yeast cells). One real-world example of cloning and expression systems is recombinant human growth hormone (hGH) production. The gene encoding hGH is inserted into a plasmid or viral vector during this process. The plasmid or viral vector is then introduced into host cells, specifically *E. coli* cells.[62] These cells can produce large quantities of the hGH protein. The cloned gene is expressed within the host cells, allowing them to produce and secrete recombinant hGH. This method enables the mass production of hGH for therapeutic purposes, such as treating growth hormone deficiencies in children or adults.

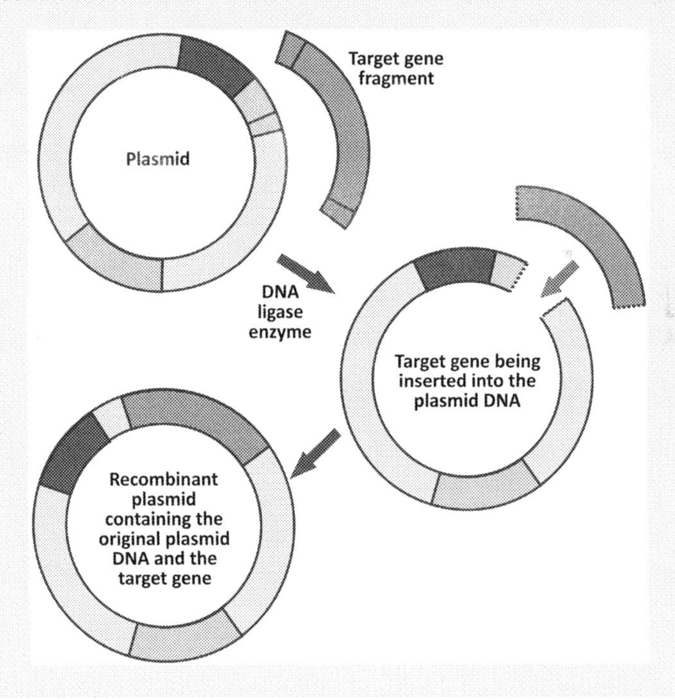

- **Directed Evolution**: This technique generates a vast library of genetic variants and implements selection processes to evolve proteins or organisms with sought-after properties. Directed evolution offers an efficient pathway to achieve protein designs that may be beyond the scope of our current understanding of protein structure and function.
  - Significant advancements in the techniques and technologies associated with directed evolution have increased its efficacy and accessibility. Developments in high-throughput screening and selection methods have enabled researchers to evaluate more extensive libraries of variants,

increasing the chance of finding a variant with the desired characteristics. Similarly, DNA synthesis and sequencing improvements have facilitated the generation and analysis of these variant libraries.

- Alongside these technological improvements, the cost of implementing directed evolution has significantly decreased, making the creation and analysis of large genetic variant libraries more affordable.
- These advancements have made directed evolution more accessible to a broader range of scientists, even those in less well-resourced settings. As a result, the technique is increasingly being utilized in various areas of synthetic biology, ranging from designing novel enzymes for industrial processes to developing new therapeutics in medicine. By making it possible to harness the power of evolution in the lab, these advancements are empowering scientists to tackle complex biological design problems and are accelerating innovation in the field.

For example, Frances Arnold, who won the 2018 Nobel Prize in Chemistry, used directed evolution to engineer enzymes. She has used these methods to create enzymes that can perform reactions not seen in nature, with applications in producing pharmaceuticals, biofuels, and more.[63]

- **DNA Sequencing**: DNA sequencing is a technique used to determine the exact order of nucleotides within a DNA molecule. It gives scientists a meticulous understanding of genetic information within organisms, enabling them to comprehend genes' functions and interactions. Crucial advancements in DNA sequencing technologies, such as deep sequencing and whole genome sequencing, have revolutionized the field of synthetic biology.
  - *Deep Sequencing*, also known as next-generation sequencing, allows scientists to sequence DNA and RNA more quickly and affordably than older sequencing techniques. This powerful tool can generate a massive amount of data, sequencing entire genomes or targeted regions of the DNA in one experiment.
  - *Whole Genome Sequencing* (*WGS*) is a comprehensive method for analyzing entire genomes. It provides a high-resolution, base-by-base view of the entire genome, offering the highest level of information content and the most comprehensive variant detection capabilities. WGS gives researchers a complete overview of all genetic variations, not just the ones occurring in known regions of the DNA.

These advancements and decreasing costs have democratized access to these potent technologies. Once the purview of large institutions and well-funded research projects, even small labs and research groups can use deep sequencing and WGS. This broad access is catalyzing rapid innovation in synthetic biology as more researchers can investigate genetic material at a granular level, opening new possibilities for understanding, manipulating, and designing biological systems.

For example, the Human Genome Project, which aimed to map and understand all the genes of human beings, relied heavily on DNA sequencing technology. WGS was used to decode the entire human genome, leading to a better understanding of many genetic diseases.[4]

- **DNA Synthesis**: A critical tool for synthetic biologists, DNA synthesis technology allows researchers to construct custom DNA sequences containing genes or genetic elements of interest for their studies or engineering applications. It enables scientists to design and fabricate genetic sequences that may not exist in nature, thereby pushing the boundaries of what is biologically possible.
    - Advancements in DNA synthesis technology, including techniques like oligonucleotide synthesis and gene synthesis, have become increasingly precise, reliable, and high throughput over the years. These improvements have enabled more complex and ambitious synthetic biology projects, fueling advancements in biofuel production, disease modeling, and personalized medicine.
    - Just as important as these technical advancements have been the dramatic decrease in the cost of DNA synthesis. This price reduction, combined with the improved accessibility of this technology, means that more researchers can design and create custom DNA sequences for their research, even in smaller or less well-funded labs. This increased access to DNA synthesis is accelerating innovation in synthetic biology, enabling a broader range of scientists to contribute to the field and facilitating the pursuit of a more comprehensive array of research questions and applications.

  For example, synthetic vaccines, such as the COVID-19 mRNA vaccines developed by Moderna and Pfizer-BioNTech, rely on DNA synthesis technology to create a DNA template of the mRNA fragment of the spike protein found on the surface of the SARS-CoV-2 virus. The vaccine contains a copy of this mRNA fragment, and when injected into the muscle, it triggers an immune response in the human body.[64]

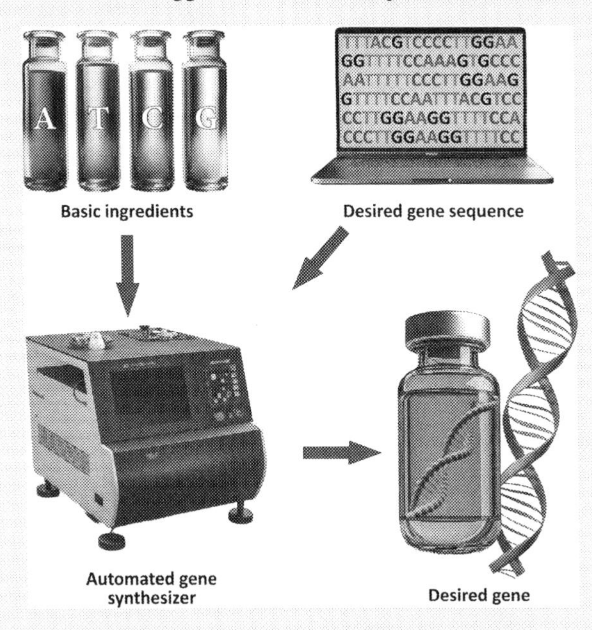

- **Gene Synthesis**: This technique involves the artificial creation of DNA sequences that can be inserted into living organisms, providing a means for designing entirely novel genetic information. Gene synthesis is integral in synthetic biology, facilitating the design and construction of new biological parts, devices, and systems.
  - Recent advancements have made gene synthesis more efficient, accurate, and scalable. High-throughput methods can now synthesize thousands of different DNA sequences simultaneously, while error-correction technologies have improved the accuracy of the synthesized sequences. Additionally, the ability to synthesize longer pieces of DNA has enabled the construction of more complex genetic systems, including entire synthetic genomes.
  - Alongside these technological advancements, gene synthesis costs have dramatically decreased over the past decade. This decrease is due to technological improvements and increased competition among gene synthesis providers. As a result, custom-made DNA sequences, which were once prohibitively expensive for many researchers, have become a common tool in many biology labs.
  - These developments have democratized access to gene synthesis, making this powerful tool available to a broader range of scientists and engineers. This wider accessibility accelerates innovation in synthetic biology, enabling more researchers to design and create custom genetic systems for various applications, from basic research to industrial biotechnology and healthcare.

  The creation of "Synthia," the first synthetic bacterial cell, by the J. Craig Venter Institute is a significant example. Scientists synthesized the entire genome of a bacterium and inserted it into a living cell, creating an entirely new organism.[20]
- **Genome Editing Tools**: CRISPR-Cas9, ZFNs (Zinc Finger Nucleases), and TALENs (Transcription Activator-Like Effector Nucleases) empower scientists to edit genomes with remarkable precision. They can add, remove, or alter genetic material at specific locations in the genome, enabling fine-tuned modifications. Among these, CRISPR-Cas9 currently stands as the most frequently employed system for genome editing due to its simplicity, efficiency, and versatility.
  - In recent years, substantial advancements in genome editing technology have made these tools more accurate and easier to use. For instance, improvements in the design and delivery of CRISPR-Cas9 systems have minimized off-target effects and increased the range of organisms and cell types in which genome editing can be performed.
  - Simultaneously, the cost of using these genome editing tools has plummeted. CRISPR-Cas9 is known for its cost-effectiveness compared to earlier genome editing technologies. The synthesis of guide RNAs for directing CRISPR-Cas9 to its target is inexpensive, and the process can be easily scaled up or down as needed.

    - These developments have made genome editing tools more accessible to a broader range of researchers, especially those in smaller labs or resource-limited settings. As a result, an increasing number of scientists can harness the power of precise genome editing for their research, contributing to a broad spectrum of applications, from basic biology research to the development of new treatments for genetic diseases. This increased accessibility is democratizing synthetic biology, driving innovation, and speeding up the translation of research findings into real-world applications.

  For example, Sickle cell disease, a severe genetic disorder, is being treated using CRISPR-Cas9 technology to modify the genes in the patient's bone marrow cells. Early clinical trials have shown promising results.[65]
- **Microfluidics**: This technique allows for precise control and manipulation of fluids at the microscale level. It is commonly used in synthetic biology for applications such as single-cell analysis, protein or DNA synthesis, and the study of cell cultures. For example, lab-on-a-chip devices combine multiple laboratory functions on a single chip, just millimeters or centimeters in size, using microfluidics technology. This has applications in point-of-care diagnostics, where immediate results are needed.
- **Polymerase Chain Reaction (PCR)**: This prevalent laboratory technique is used to amplify specific segments of DNA, creating multiple copies from a minute starting amount. It underpins numerous genetic analyses because it empowers scientists to generate ample quantities of specific DNA segments for study.
    - Over time, advancements in PCR technology have made the process more reliable, versatile, and high throughput. For instance, the development of real-time PCR allows scientists to monitor the amplification process as it happens, providing quantitative results. Other variants of PCR, such as reverse transcription PCR (RT-PCR), enable the analysis of RNA sequences, while multiplex PCR can amplify multiple target sequences simultaneously.
    - Coupled with these advancements, PCR costs have significantly decreased in recent years. This reduction is due to the increased efficiency of the process, cheaper and more robust thermal cyclers, and the reduced cost of PCR reagents.
    - These developments have democratized access to PCR, making it a standard and affordable tool in all biology labs worldwide. This widespread accessibility is catalyzing progress in synthetic biology, enabling more researchers to analyze genetic material and contributing to a myriad of applications, from basic genetic research to diagnostic testing, forensics, and beyond.

  For example, PCR has been widely used during the COVID-19 pandemic for diagnostic testing. It amplifies the viral RNA present in patient samples, allowing for detection even when the viral load is low.
- **Synthetic Gene Networks**: These are designed and built as systems of genes that interact in a cell to produce the desired behavior. Designing these

networks involves principles from electrical engineering and computer science. The creation of a biological toggle switch in *E. coli* bacteria is an example. The switch is a simple synthetic genetic network that contains two repressor proteins, each of which inhibits the production of the other, creating a bistable system that can be "flipped" by external signals.[37]

These tools and techniques form the foundation of synthetic biology, allowing scientists across all disciplines to manipulate biological systems in unprecedented ways.

2.3 SYNTHETIC TISSUES AND ORGANS

Synthetic biology holds immense potential in the field of tissue engineering and regenerative medicine to replace or regenerate human cells, tissues, or organs to restore or establish normal function. Ways how synthetic biology is contributing to the development of synthetic tissues and organs include:

- **Bioprinting**: This process uses 3D printing technology, but instead of using plastics or metals, it uses bioinks made up of cells and other biological materials. The 3D bioprinter is used to create complex tissue structures layer by layer that mimic natural tissues. Bioprinting can create tissues for transplantation or for *in vitro* testing of drugs or other substances. An example of bioprinting is the work done by researchers at the Wake Forest Institute for Regenerative Medicine. They have successfully bioprinted ear, bone, and muscle structures, which, once implanted in animals, matured into functional tissue and developed a system of blood vessels.[66]

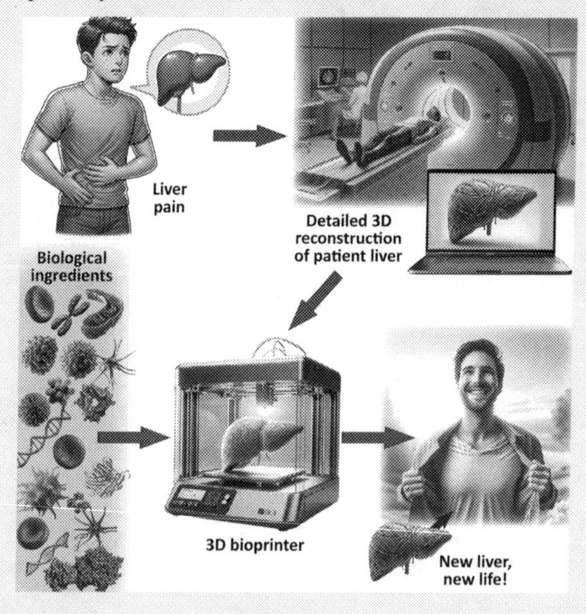

- **Biomaterials**: Synthetic biology can design new biomaterials or modify existing ones for use in tissue engineering. These biomaterials can be designed to interact with cells in specific ways, guiding tissue formation. For example, they can be engineered to release growth factors that stimulate cell growth or to have mechanical properties that mimic the natural tissue. One example of a biomaterial designed through synthetic biology is the development of hydrogels, which are often used in tissue engineering. Hydrogels can be designed to have specific properties such as the controlled release of growth factors, as seen in a study by researchers at MIT where they developed a hydrogel that releases drugs in response to mechanical force to aid wound healing.[67]
- **Organoids**: Organoids are miniature, simplified versions of organs produced *in vitro* in three dimensions that show realistic micro-anatomy. They are derived from stem cells and have realistic micro-anatomy, making them useful for studying organ development, disease, and drug response. Organoids have been used in various fields of medical research. For instance, scientists have created mini-brains or brain organoids to study the development and disorders of the human brain, including diseases like Alzheimer's and Parkinson's.[68]
- **Organ-on-a-Chip Technology**: This technology involves creating miniaturized versions of human organs on microfluidic chips. Each organ-on-a-chip is made up of a clear, flexible polymer that contains hollow microchannels lined with living human cells. These chips can mimic the mechanical and physiological response of whole organs. The Wyss Institute at Harvard University has developed multiple organ-on-a-chip models, including lung-on-a-chip[69] and heart-on-a-chip.[70] These models test drug responses and study disease mechanisms in a more realistic, human-relevant environment.
- **Decellularized Organs**: This involves removing all the cells from a donor organ, leaving behind a collagen scaffold. This scaffold, which maintains the intricate architecture of the original organ, can then be repopulated with the patient's own cells, reducing the risk of organ rejection. An example of this technology was demonstrated by researchers at Massachusetts General Hospital and Harvard Medical School. They successfully decellularized a rat kidney and then recellularized it with new cells,[71] after which the kidney produced urine in a way similar to a naturally grown kidney.

While considerable progress has been made, there are still many challenges to overcome, including technical issues related to cell sourcing, organ vascularization, organ maturation, and maintaining cells' viability and function. However, the potential of synthetic biology for tissue engineering and regenerative medicine is enormous and could revolutionize healthcare in the future.

2.4 PERSONALIZED MEDICINE

Personalized medicine seeks to tailor medical treatments to individuals based on their unique genetic makeup, physiological status, and environmental influences. This approach to healthcare is expected to improve efficacy and reduce side effects. Synthetic biology plays a pivotal role in advancing personalized medicine.

One real-world example is the case of cancer chemotherapy, where the therapeutic window is narrow and side effects can be severe. Personalized medicine can significantly improve treatment outcomes. Using synthetic biology, scientists can identify specific genes that dictate drug response in cancer patients. By analyzing the patient's genetic profile, doctors can prescribe drugs most effective for their specific cancer type and genetic variations, while minimizing adverse reactions. This personalized approach allows for tailored treatments that maximize efficacy and minimize side effects.

Synthetic biology is a powerful tool in the evolution of personalized medicine, allowing treatments to be tailored to the specific genetic and physiological characteristics of each individual. While much of this is still in the research phase, the potential for transformative changes in healthcare is immense.

2.5 OTHER ADVANCEMENTS

- **Gene Drive Technology**: Gene drive has the potential to alter the genetic makeup of entire populations or even eradicate species, such as disease-spreading mosquitoes. However, this technology poses significant biosecurity risks, including the potential for accidental release, misuse, or the unforeseen ecological consequences of altering ecosystems. An example of this technology is the work conducted exploring gene drive technology to combat malaria by reducing the population of malaria-spreading mosquitoes.[22] The researchers aim to modify the genes of the mosquitoes so that they produce almost entirely male offspring, reducing the population over time as fewer females are available to reproduce. This example showcases the great potential gene drive has to help some of the worst public health problems we have. Yet, it also underlines the potential biosecurity risks associated with gene drive technology, such as the unforeseen ecological consequences of significantly reducing a species.
- **DNA Synthesis Screening**: Advances in DNA synthesis technology have made it easier and more affordable to create custom DNA sequences. This has numerous beneficial applications, but there are concerns regarding the potential synthesis of harmful or weaponized biological agents. Improved DNA synthesis screening methods are being developed to detect and prevent the synthesis of such sequences, enhancing biosecurity. One example of this is the synthesis of the horsepox virus by researchers in Canada in 2017.[72] They assembled the virus using pieces of DNA purchased from a commercial supplier, highlighting the potential risks associated with the widespread accessibility of DNA synthesis technology. To mitigate such risks, companies like Integrated DNA Technologies screen orders for sequences related to hazardous pathogens, aiming to prevent the misuse of synthetic biology.
- **CRISPR-Based Diagnostics**: CRISPR technology is revolutionizing diagnostics for the rapid, accurate, and portable detection of pathogens. This has substantial benefits for public health, but it also raises biosecurity concerns, as the knowledge and tools required to edit genomes could be misused. Addressing these concerns involves developing ethical guidelines and regulatory frameworks around the use of CRISPR. An example is the development

of the SHERLOCK (Specific High-sensitivity Enzymatic Reporter unLOCKing) and DETECTR (DNA Endonuclease-Targeted CRISPR Trans Reporter) systems.[73] These CRISPR-based tools can identify specific genetic sequences, making them useful for the rapid detection of pathogens like Zika virus or SARS-CoV-2. However, the same knowledge and tools could potentially create harmful genetic modifications, raising biosecurity concerns.

- **De-extinction and Synthetic Organisms**: Synthetic biology has opened the door for the potential resurrection of extinct species or the creation of entirely new organisms. These advances have exciting implications for biodiversity, conservation, and biomedicine. They also pose biosecurity risks due to unforeseen ecological impacts, potential misuse, and ethical considerations. One example is the effort to bring back the woolly mammoth, an extinct species, by a team of geneticists at Harvard University.[74] By introducing genes from the woolly mammoth into the genome of an Asian elephant (the mammoth's closest living relative), they hope to create a hybrid that can survive in the Arctic. This project underscores the potential biosecurity risks of such technology, including unforeseen ecological impacts and ethical considerations.
- **AI-Driven Tools and Platforms**: Artificial Intelligence platforms and machine learning are tremendously accelerating progress in synthetic biology research and its applications. These advances promise more efficient and accurate results, but they also raise biosecurity concerns, such as the potential for automated biohacking or the creation of dangerous synthetic organisms. Addressing these concerns requires interdisciplinary collaboration between AI engineers and synthetic biology experts to develop safeguards and ethical guidelines. For further information, refer to Chapter 4, *The Role of Artificial Intelligence (AI) in Synthetic Biology and Biosecurity Implications.*

2.6 CONCLUDING THOUGHTS

Synthetic biology's profound impact on biomedicine is catalyzing a paradigm shift in healthcare and medical research. This field is transforming the landscape of biomedicine through diverse applications such as drug discovery, personalized medicine, and the creation of synthetic tissues and organs. The ability to redesign and construct biological systems for specific medical purposes brings a new era of breakthrough treatments and diagnostic tools. However, addressing ethical, biosafety, and biosecurity concerns becomes increasingly critical as we embrace these advancements. Developing robust regulatory frameworks to ensure safe and responsible application is as crucial as the innovations. We are on the verge of a future where synthetic biology, coupled with meticulous regulation and ethical considerations, could revolutionize healthcare while significantly enhancing human health and well-being. The potential to customize medical treatments to individual genetic profiles, engineer tissues and organs, and develop novel therapeutic agents is extraordinary. Nonetheless, these advancements' biosecurity risks must be considered. The scientific community, policymakers, biorisk managing professionals, and the public must engage in ongoing dialogue to navigate these complex issues. Balancing innovation with biorisk management will be vital to harnessing the full potential of synthetic biology in improving human health, ensuring that advancements in biomedicine are both groundbreaking and ethically sound.

2.7 KEY TAKEAWAYS

- **Defining Synthetic Biology**: An interdisciplinary field applying engineering principles to biology, aiming to design and construct new biological systems.
- **Applications in Biomedicine**: Include drug discovery, personalized medicine, tissue engineering, and biosynthetic drugs and vaccines.
- **Ethical and Safety Concerns**: Address biosafety, biosecurity, and ethical implications of genetic manipulation.
- **Regulatory Frameworks**: Emphasize the need for robust regulations for safe and responsible development.
- **Future Directions**: Anticipate synthetic biology's potential to further revolutionize healthcare.

2.8 THOUGHT-PROVOKING QUESTIONS

2.1 Why is synthetic biology's safe and responsible development, deployment, and application important to advancing this field?

2.2 What would be the most critical components of a comprehensive plan to educate the public about the fundamentals of synthetic biology?

2.3 Why is understanding the fundamentals of synthetic biology crucial for policymakers?

2.4 How can biosecurity be ensured in a world where synthetic biology technologies are becoming increasingly accessible?

2.5 What are the challenges of synthetic biology in tissue engineering and regenerative medicine?

2.6 What are the potential ethical implications for the most recent innovative advancement of synthetic biology applied to personalized medicine?

2.7 How can the biosecurity risks associated with gene drive technology and DNA synthesis screening be mitigated?

2.8 What are the potential biosecurity risks associated with experiments resurrecting extinct species or creating entirely new organisms through synthetic biology? Do the benefits outweigh the risks?

2.9 Perform a risk-benefit analysis for the potential environmental impacts of using microorganisms for drug production. Do the benefits outweigh the risks? How can the risks be mitigated?

2.10 How can the risk of misuse of synthetic biology in drug production, such as the production of opioids, be mitigated?

2.11 What are the ethical implications of using engineered microorganisms for drug production, and how can they be addressed?

2.12 How can synthetic biology contribute to reducing the burden on healthcare systems and improving global health?

3 Biosecurity Risks in Synthetic Biology

Biorisk management involves a structured approach to identifying, evaluating, and reducing risks linked to the use, handling, and storage of biological agents and toxins. This process aims to safeguard individuals, the broader public, and the environment from potential dangers posed by biological materials. It incorporates two key elements: biosafety, which focuses on preventing accidental exposure or release of pathogens and toxins, and biosecurity, which concentrates on preventing unauthorized access, theft, or misuse of these materials. Together, these two elements form a comprehensive biorisk management framework. In this chapter, we will explore biosecurity in greater detail, particularly in the context of synthetic biology. We will discuss specific risks associated with this field, illustrating them with tangible examples. The chapter aims to provide a clear understanding of the unique biosecurity challenges in synthetic biology and offer practical insights into managing these risks effectively.

DOI: 10.1201/9781003423171-3

3.1 BIOSAFETY & BIOSECURITY: DIFFERENCES AND COMMONALITIES

In the rapidly advancing field of synthetic biology, ensuring the safe and responsible development, deployment, and application of novel biological systems is important. Two key areas that address the management and mitigation of potential risks associated with synthetic biology are biosafety and biosecurity. While biosafety and biosecurity share a common goal of safeguarding public health and the environment, they encompass distinct concepts and approaches, as illustrated in Figure 3.1. This section aims to provide an extensive definition of biosafety and biosecurity, elucidate their differences, and highlight their underlying commonalities.

- **Definition of Biosafety**: Biosafety can be defined as the discipline focused on preventing or minimizing harm resulting from the unintentional release of biological agents or genetically modified organisms (GMOs) into the environment. It encompasses practices, protocols, and strategies to reduce the risks associated with handling, storing, and disposing hazardous biological materials. Biosafety measures protect individuals working with biological agents, the broader community, and ecosystems from potential adverse effects on health and the environment. An example of this is the handling

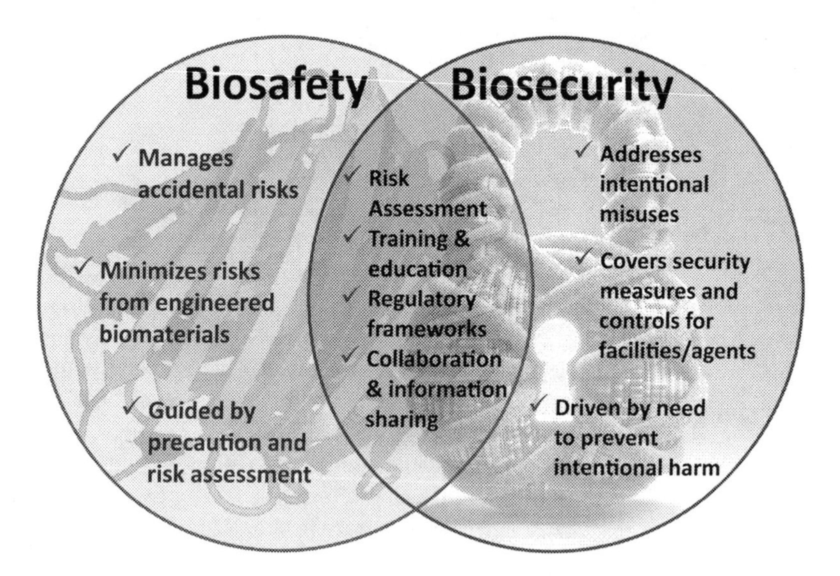

FIGURE 3.1 Biosafety & Biosecurity: differences and commonalities. Both biosafety and biosecurity form integral parts of a comprehensive biorisk management framework. These two elements share several common areas, including risk assessment, education, training, regulatory adherence, and sharing information and collaboration. However, they distinctly differ in their key focus areas: Biosafety is primarily concerned with protecting individuals and the environment from potential harm arising from biological materials or organisms. At the same time, Biosecurity emphasizes the protection of valuable biological materials (VBMs) from deliberate theft or misuse.

of the Ebola virus in research laboratories. Given the virus's high lethality and ease of transmission, strict biosafety measures are implemented to protect lab workers and the surrounding environment. These measures include wearing personal protective equipment (PPE), using maximum containment measures, often referred to as Biosafety Level 4 facilities (the highest level of biological containment – see Section 5.1.4 for a detailed discussion), and adhering to strict decontamination and waste disposal protocols.

- **Definition of Biosecurity**: The term "biosecurity" originally described precautions taken to stop harmful organisms from entering the environment, particularly to safeguard agricultural crops and livestock.[75–77] However, its meaning has broadened to encompass a wide array of strategies aimed at deterring the intentional misuse, theft, loss, or unauthorized access to biological agents, pathogens, toxins, or related facilities, known as Lab Biosecurity. In this book, "biosecurity" will specifically refer to Lab Biosecurity, unless otherwise specified. Biosecurity encompasses protecting biological resources, facilities, and information from deliberate acts of misuse, including bioterrorism. Biosecurity measures involve implementing stringent physical, operational, and personnel security measures to prevent unauthorized access or intentional release of dangerous biological materials, thereby reducing the risks posed by the deliberate use of biological agents for harmful purposes. A real-world example is the U.S. regulation and procedures that govern access to select agents – biological agents and toxins that potentially pose a severe threat to public health and safety, animal and plant health, or animal and plant products. In the United States, the Centers for Disease Control and Prevention (CDC) and the Animal and Plant Health Inspection Service (APHIS) regulate these agents' possession, use, and transfer, requiring labs to implement stringent biosecurity measures to prevent theft or misuse.[12]
- **Differences between Biosafety and Biosecurity**: Although biosafety and biosecurity share the objective of preventing harm, they differ in their primary focus, scope, and underlying principles:
    - **Focus**: Biosafety primarily addresses the **unintentional or accidental** release of biological agents, focusing on minimizing risks associated with laboratory and industrial practices involving GMOs or hazardous biological materials. It emphasizes risk assessment, containment, and safe handling procedures to protect laboratory workers, the environment, and public health. On the other hand, biosecurity deals with the **intentional misuse or unauthorized access** to biological agents or toxins and related facilities. It encompasses measures to prevent the intentional release or theft of biological materials. It aims to safeguard national security and public safety against individuals or groups' deliberate use of biological agents with malicious intent.
    - **Scope**: Biosafety focuses on managing and minimizing the risks associated with laboratory or industrial processes involving biological agents or GMOs. Biosecurity has a broader scope, encompassing physical security measures, personnel security, access controls, information security,

and the establishment of regulatory frameworks and international collaborations to prevent the misuse of biological agents. It extends beyond laboratory settings, including transportation, storage, and regulation of dangerous biological materials.

- **Underlying Principles**: Biosafety principles are guided by a precautionary approach and the concept of risk assessment. They aim to minimize potential harm by identifying and implementing appropriate containment and control measures, emphasizing safe laboratory practices, and promoting responsible conduct in research and development. Biosecurity principles are driven by the need to protect against intentional acts of harm and ensure the security of biological materials and facilities.

For instance, in a research lab studying influenza viruses, biosafety concerns might involve ensuring proper use of PPE, installing appropriate ventilation systems, and enforcing protocols for handling and disposing of potentially infectious materials to protect lab workers and prevent accidental release of the virus. In contrast, biosecurity concerns in the same lab might include controlling access to areas where the viruses are stored, conducting background checks on lab personnel, and setting up procedures to ensure that viruses are not removed from the lab and misused.

- **Commonalities between Biosafety and Biosecurity**: While biosafety and biosecurity differ in focus and scope, they share several commonalities:
  - **Risk Assessment**: Both biosafety and biosecurity employ risk assessment methodologies to evaluate the potential risks associated with biological agents, GMOs, or pathogens. This includes identifying hazards, assessing the likelihood of harm, and implementing appropriate control measures to mitigate risks.
  - **Training and Education**: Both biosafety and biosecurity emphasize the importance of training and education to promote awareness, knowledge, and adherence to safe practices. This includes training on proper handling, disposal, and containment procedures, as well as educating personnel about the risks associated with intentional misuse.
  - **Regulatory Frameworks**: Biosafety and biosecurity rely on robust regulatory frameworks to ensure safety and security standards compliance. These frameworks encompass guidelines, protocols, and regulations that govern the handling, transportation, and storage of biological agents, GMOs, or hazardous materials.
  - **Collaboration and Information Sharing**: Biosafety and biosecurity require collaboration and information sharing among researchers, institutions, and governmental bodies at national and international levels. This collaboration facilitates developing best practices, harmonizing standards, and exchanging information related to emerging risks and mitigation strategies.

The anthrax attacks in the United States 2001 highlight these commonalities. In response to the attacks, labs handling anthrax and other select agents had to implement biosafety measures (to prevent accidental exposure or release of the bacteria) and biosecurity measures (to prevent theft or misuse of the bacteria). Risk assessments were used to guide these measures, personnel received training on biosafety and biosecurity, and regulatory frameworks were updated to enforce compliance. Additionally, these events underscored the importance of collaboration and information sharing between labs, law enforcement, and public health agencies, both within the United States and internationally.[12]

Biosafety and biosecurity play crucial roles in ensuring synthetic biology's safe and responsible advancement. While biosafety addresses unintentional harm associated with biological agents and GMOs, biosecurity focuses on intentional acts of misuse. Despite their differences, both fields share commonalities in risk assessment, training and education, regulatory frameworks, and the need for collaboration and information sharing. By integrating biosafety and biosecurity measures, stakeholders can collectively mitigate risks, foster responsible innovation, and ensure the long-term sustainability of synthetic biology.

3.2 POTENTIAL MISUSE AND DUAL-USE CONCERNS

The promise and power of synthetic biology also bring significant potential risks. Just as the field offers opportunities to improve health, protect the environment, and build new industries, it also has the potential to be misused, either intentionally or unintentionally. Some of the potential misuse and dual-use concerns related to synthetic biology include:

- **Bioweapons**: Synthetic biology could theoretically create harmful biological agents or toxins. For instance, someone with malicious intent could use synthetic biology to engineer a pathogen that is more deadly, more resistant to current treatments, or can target specific groups of people. It could also potentially be used to recreate known pathogens from scratch.
- **Bioterrorism**: Synthetic biology could be used in bioterrorism, for instance, by releasing engineered organisms into the environment or food supply. These organisms could be designed to produce toxins or to cause damage to crops or ecosystems.
- **Unintended Consequences**: Even with good intentions, releasing synthetic organisms into the environment could have unintended and potentially catastrophic consequences. These organisms could outcompete natural organisms, disrupt ecosystems, or transfer their engineered genes to natural organisms.
- **Biohacking**: The democratization of biology, with the advent of DIY bio labs and citizen science, raises concerns about the potential for misuse. While most of these community scientists are working on benign projects, there is the risk that some could use their skills and access to equipment for malicious purposes.
- **Ethical Concerns**: The ability to manipulate life at the most fundamental level raises profound ethical questions. For instance, where should the line be drawn between therapy and enhancement? What are the implications of creating entirely synthetic organisms?

- **Equity Concerns**: Like many technologies, the benefits of synthetic biology are likely to be available first to those who can afford them. This raises concerns about exacerbating existing inequities in health, well-being, and the environment.

To mitigate these risks, experts have called for a combination of technical measures (like building safety features into engineered organisms), regulatory measures (like oversight of research and commercial activities), and the development of norms and codes of conduct for practitioners of synthetic biology. They also stress the importance of transparency and public engagement to build trust and ensure that the development of synthetic biology aligns with societal values.

3.2.1 Dual-Use Research of Concern (DURC)

- **Dual-Use Technology**: Synthetic biology tools and techniques have the potential for both beneficial and harmful applications. The same technologies for developing life-saving therapies or sustainable solutions could also be misused for malevolent purposes. Promoting responsible research and ensuring ethical guidelines and oversight are critical to minimizing the dual-use risks. For example, the reconstruction of the deadly 1918 influenza virus from preserved genetic material demonstrates how synthetic biology capabilities could resurrect extinct pathogens, highlighting the need for careful oversight of such research.[78]

3.2.2 The Potential Misuse of Synthetic Biology

- **Accidental Release of Engineered Organisms**: The accidental release of engineered organisms is a significant concern in synthetic biology due to the potential for unintended consequences. Despite strict containment measures in laboratories, there is always the risk of an unforeseen event leading to the escape of these organisms into the environment. Several scenarios could unfold with the accidental release of engineered organisms:
  - **Disruption of Ecosystems**: Genetically engineered organisms could interfere with existing ecosystems in unpredictable ways. They might outcompete native species, disrupt food chains, or introduce new diseases. These effects could lead to the loss of biodiversity and the disruption of ecosystem services.
  - **Horizontal Gene Transfer**: Engineered organisms might pass on their synthetic genes to other organisms through a process known as horizontal gene transfer. This could lead to the spread of novel traits in non-target species, which could have unforeseen ecological impacts.
  - **Evolution and Adaptation**: Over time, engineered organisms could evolve and adapt in ways that were not initially predicted, potentially resulting in harmful traits or behaviors. Additionally, if the organism is designed to replicate, mutations could occur over generations leading to unexpected changes.

   - **Human Health Risks**: If engineered organisms were to come into contact with humans, there could be potential health risks. For example, engineered microbes might cause new diseases or disrupt the human microbiome.
   - **Economic Impact**: The accidental release of an engineered organism could have significant economic impacts, particularly if it affects agriculture, forestry, fisheries, or other industries that rely on natural resources.

  To minimize these risks, synthetic biologists use various containment strategies, such as physical containment measures in labs, and biological containment methods, such as designing organisms to be unable to survive or reproduce outside of the lab. However, these measures are not 100% foolproof, and there is an ongoing debate about how to manage the potential risks of accidental release and what mitigation measures should be put in place. Developing robust risk assessment frameworks, regulatory oversight, and emergency response plans are critical for mitigating these risks.

- **Intentional Release of Engineered Organisms**: Intentional release of engineered organisms can pose significant biosecurity risks, particularly if such release is malicious. A few possible scenarios include:
   - **Biological Warfare or Bioterrorism**: Synthetic biology could theoretically be used to engineer harmful biological agents for use in biological warfare or bioterrorism. This could include creating new pathogens, making existing pathogens more virulent or resistant to treatment, or designing organisms that produce toxins or other harmful substances. For example, a malicious actor might engineer a virus to be more contagious or deadly, and then intentionally release it.
   - **Sabotage of Agriculture or Environment**: Engineered organisms could be used to harm agriculture or the environment. For instance, bacteria could be designed to attack a specific crop, causing widespread crop failure and economic damage. Similarly, an engineered organism could harm specific species or disrupt ecosystems.
   - **Disruption of Microbiomes**: An engineered microorganism could be intentionally released to disrupt human or animal microbiomes. The human microbiome, for example, plays crucial roles in digestion, immune function, and other processes, and disruption could lead to health problems.

  To prevent these scenarios, it is important to have strong biosecurity measures. This includes physical security measures to prevent unauthorized access to labs and biological materials, as well as regulatory measures, such as restrictions on certain types of research or the transport of certain materials. In addition, it is important to have surveillance systems to detect the intentional release of engineered organisms, and response plans to mitigate the effects of such an event.

  Education and culture also play a key role. By fostering a culture of responsibility in the scientific community and by educating scientists and the public about the potential risks of synthetic biology, we can help to prevent misuse of this powerful technology.

3.2.3 Intellectual Property (IP) and Data Security

- **Safeguarding Proprietary Information & Technology**: Safeguarding proprietary information and technology is a significant concern in synthetic biology, just as in other areas of technology and innovation. Some considerations for managing these issues include:
  - **Patents and Trademarks**: Patenting is a common way to protect intellectual property (IP). It grants the patent holder exclusive rights to an invention for a certain period of time, typically 20 years. Trademarks can protect brand names or logos associated with a product or service. However, navigating the patent landscape in synthetic biology can be challenging given the intersection of biological systems and engineering principles.
  - **Trade Secrets**: In some cases, companies may opt to keep their inventions as trade secrets, especially if the process of obtaining a patent might disclose valuable information to competitors. This means ensuring that information is kept confidential and only accessible to those who need to know.
  - **Cybersecurity Measures**: As much synthetic biology involves digital data, robust cybersecurity measures are necessary to protect proprietary information from being stolen or tampered with. This includes using secure networks, implementing firewalls, regularly updating and patching systems, and training staff in cybersecurity best practices.
  - **Non-Disclosure Agreements (NDAs)**: NDAs can ensure that employees, contractors, and collaborators do not disclose sensitive information. These legally binding contracts can help protect IP and proprietary technology.
  - **Physical Security**: Physical security measures are important for protecting sensitive materials and equipment, such as lab notebooks, biological samples, and specialized lab equipment.
  - **Due Diligence in Collaborations and Partnerships**: When entering collaborations or partnerships, it is important to conduct due diligence to ensure that all parties will appropriately protect proprietary information and respect IP rights.

Balancing IP protection with the ethos of openness and collaboration in science can be challenging. Openness is key to scientific progress, but without IP protection, companies may lack the incentive to invest in the expensive and time-consuming process of developing new products and technologies. Therefore, finding a balance that encourages innovation and collaboration is key to the future of synthetic biology.

Example: German biotechnology company Evotec recently experienced a major cybersecurity breach that impacted its proprietary research data and IP.[79]

In early April 2022, Evotec detected unusual activity on its IT systems, prompting the company to immediately take its entire digital infrastructure offline. Further investigation revealed that Evotec had been the victim of a sophisticated cyberattack perpetrated by unknown threat actors who infiltrated their network and exfiltrated sensitive files. This real-world case aligns with the increasing cyber threat trends faced by biopharmaceutical firms safeguarding valuable IP and underscores the importance of robust, multilayered data security to prevent catastrophic breaches. As biotech research grows ever more complex and data-driven, proactive governance, risk management, access controls, infrastructure security, encryption, and staff training are essential to defend against criminal cyber groups looking to profit from stolen innovation.

- **The Risk of Bioterrorism through Information Theft**: The integration of biology with information technology in synthetic biology means that there are new kinds of biosecurity risks to consider. Among these is the risk of bioterrorism facilitated by information theft. Some ways this could occur include:
    - **Stealing Genetic Data**: Genetic data, such as the genome sequences of organisms, are crucial in synthetic biology. If such data were stolen, it could potentially be used to synthesize harmful organisms, especially if the data pertains to pathogenic viruses or bacteria.
    - **Theft of Bioengineering Techniques**: Synthetic biology relies on specific bioengineering techniques and protocols. If detailed information about these techniques were stolen, it could be used by malicious actors to engineer harmful organisms or biological materials.
    - **Cyberattacks on Bioinformatics Systems**: Many synthetic biology labs rely heavily on computational systems for tasks such as designing genetic constructs or modeling biological systems. If these systems were compromised in a cyberattack, it could potentially allow malicious actors to gain access to sensitive information or disrupt crucial research activities.
    - **Misuse of Openly Available Information**: While not strictly "theft," there is also a concern that openly available information in synthetic biology could be misused. One example is an experiment by U.S. scientists to demonstrate that the polio virus can be created from scratch using data from publicly available publications. The experiment showed that viruses can be produced from scratch, including smallpox and Ebola, that can be used to manufacture viruses for bioterrorism.[33]

Mitigating these risks requires robust data security measures. This includes physical security measures, such as securing labs and storage facilities, and digital security measures, such as encryption, secure data storage and transfer protocols, and strong access controls. It also includes legal and policy measures, such as laws against biohacking and misuse of biological information, and guidelines for what information should be openly shared vs. kept confidential.[80]

Education and awareness are also crucial. Researchers, technicians, and students in synthetic biology need to be aware of the potential security implications of their work and be trained in best practices for data security. Regular audits and updates of security measures can help ensure that they remain effective as technologies and threats evolve.

Example: A real-world example that illustrates the concepts of the risk of bioterrorism through information theft is that data breaches at research institutions and universities are increasing every year, to the point that it is being described as a crisis by experts in the field.[81] Universities conducting sensitive synthetic biology research could have their systems compromised, allowing theft of genomic data or bioengineering techniques. If vulnerabilities are left unaddressed, motivated hackers could gain access to confidential information and repurpose it for bioterrorism. The consequences of such data falling into the wrong hands could be dire. Malicious actors could synthesize dangerous pathogens, enhance the lethality of biological weapons, or sabotage critical research efforts. Simply doing nothing to improve data security practices magnifies the risk that sensitive information will be misused for catastrophic ends. Proactive measures like access controls, personnel training, and robust cybersecurity defenses are crucial to mitigate this threat. The rising frequency of university data breaches underscores the need to be vigilant and proactive when securing potentially hazardous biological information.

3.3 THREATS POSED BY ENGINEERED PATHOGENS

Engineered pathogens are a significant concern when discussing the potential risks of synthetic biology. While there are numerous safeguards and ethical guidelines for bioengineering research, the risk of engineered pathogens still exists due to several factors:

- **Creating More Dangerous Pathogens**: With the tools of synthetic biology, it is theoretically possible to alter existing pathogens to make them more deadly, more easily transmissible, or more resistant to existing treatments. Scientists might do this with good intentions, such as wanting to study the pathogen to develop better treatments. However, there is a risk that such a pathogen could accidentally escape from the lab.
- **Recreating Known Pathogens**: Synthetic biology also makes it possible to synthesize known pathogens from scratch using publicly available genetic sequences. For instance, in 2005, researchers in the United States recreated the 1918 Spanish flu virus to study its properties. This raised concerns that someone with malicious intent could recreate this deadly virus.[82]
- **Creating Novel Pathogens**: Theoretically, it might also be possible to create entirely novel pathogens that the immune system does not recognize and cannot fight off.
- **Weaponizing Pathogens**: In the wrong hands, synthetic biology could be used to weaponize pathogens. This could involve making the pathogen more

deadly, transmissible, or capable of surviving in harsh conditions. It could also involve encoding the pathogen to target specific populations based on genetic markers – a grim prospect known as a "genetic weapon."

To mitigate these risks, a number of measures must be implemented, which include:

- **Safety and Security Measures**: Laboratories working with dangerous pathogens must follow strict safety and security protocols to prevent accidental release. However, these measures need to be constantly updated as synthetic biology evolves.
- **Regulation and Oversight**: National and international regulations and oversight mechanisms are required to monitor research involving potential pandemic pathogens and to prevent the misuse of synthetic biology.
- **Transparency and Responsible Communication**: While it is important for researchers to share their findings, caution is needed to prevent the misuse of potentially dangerous information. There is an ongoing debate among scientists about how to balance the need for openness with the need for security.
- **Public Engagement**: The public should be involved in discussions about the benefits and risks of synthetic biology, as well as how to manage these risks. This can help to build trust and ensure that the development of synthetic biology aligns with societal values.
- **Education and Ethics**: Training in biosecurity and bioethics should be an essential part of the education of all synthetic biologists.

Despite these potential risks, it is important to note that synthetic biology also offers tremendous potential benefits, including new ways to treat diseases, produce food and energy, and protect the environment. The challenge is to harness these benefits while effectively managing the risks.

3.4 THE RISK OF ORGANISMS OR GENETIC TRAITS ESCAPING

Synthetic biology's capacity to engineer organisms or genetic traits brings immense potential benefits, such as new drugs, biofuels, or means of carbon capture. However, alongside these benefits, there are risks associated with the possible escape or unintended release of these engineered organisms or their genetic traits into the environment. Some key concerns include:

- **Unintended Ecosystem Disruption**: If engineered organisms were to escape into the environment, they could potentially disrupt local ecosystems. Depending on their characteristics, they might outcompete or interbreed with natural organisms, leading to unforeseen consequences for biodiversity, food chains, or other ecosystem functions.[15]
- **Spread of Engineered Traits**: Genetically engineered traits might not stay confined to the organisms they were originally introduced into. Through processes like horizontal gene transfer, these traits could potentially spread

to other organisms in the environment, leading to unpredictable ecological impacts.

- **Evolutionary Changes**: Over time, engineered organisms and their descendants could evolve in ways that are hard to predict. This could potentially lead to new pathogens or other harmful organisms.
- **Health Risks**: Depending on the nature of the engineered organisms, there could be direct health risks to humans if they escaped into the environment. For example, organisms engineered to produce pharmaceuticals or other chemicals might contaminate food or water supplies.
- **Economic Impact**: An escaped engineered organism could have negative economic impacts, for example, by damaging crops or livestock, or by disrupting ecosystem services that industries depend on.

3.4.1 Mitigation Strategies

- **Containment Strategies**: Containment strategies aim to prevent the escape of engineered organisms from laboratories or industrial facilities. These can include physical containment measures and biological containment measures, such as designing organisms that are unable to survive outside of specific controlled conditions. An example of containment strategies can be seen in labs working with GMOs, where they use specific growth mediums that these organisms require to survive, and which are not found outside the lab. This physical and biological containment prevents their survival if they escape the lab environment.
- **Control Measures**: Control measures aim to limit the spread or impact of engineered organisms in the event of an escape. These can include designing organisms with "kill switches" that can be activated to kill the organisms if necessary, or with "suicide genes" that cause the organisms to self-destruct under certain conditions. One of the practical control measures is the implementation of "kill switches" in genetically engineered bacteria. For instance, researchers have designed *Escherichia coli* bacteria that rely on synthetic amino acids not found in nature. If these bacteria escape, they cannot access the required amino acids in the natural environment, leading to their death.
- **Monitoring and Detection**: Monitoring and detection methods are also crucial for quickly identifying and responding to any escape of engineered organisms. This could involve genetic barcoding of engineered organisms to allow for their easy identification in the environment. Genetically engineered mosquitoes designed to combat the spread of diseases such as Zika or Dengue provide a good example of monitoring and detection. These mosquitoes carry a genetic barcode, allowing scientists to track their spread and measure their impact on the mosquito population in the environment.
- **Good Microbiological Practices and Procedures (GMPP)**: GMPP is a fundamental mitigation strategy in managing biosecurity risks within synthetic biology. By nature, synthetic biology involves the manipulation and creation of novel biological parts, devices, and systems, which

can inadvertently pose biosecurity risks. GMPP establishes a set of recommended procedures and practices to reduce the risks associated with handling and manipulating biological agents. This includes guidelines on laboratory hygiene, proper use and maintenance of containment facilities, PPE use, waste disposal, and incident reporting and response. The diligent application of GMPP principles not only minimizes the risks associated with the accidental release or exposure to harmful biological agents, but it also forms a key part of a comprehensive strategy for mitigating the potential misuse of synthetic biology for malicious purposes. Thus, adhering to GMPP is critical in ensuring the safe and responsible advancement of synthetic biology.

- **Regulation and Oversight**: Rigorous regulation and oversight can help to ensure that all necessary precautions are taken and that potential risks are thoroughly evaluated before any deliberate release of engineered organisms into the wild. The use of genetically engineered salmon for food consumption offers an example of regulation and oversight. The FDA evaluated the risks and benefits of these salmon before approving their use. This included ensuring that the engineered fish were as safe to eat as non-engineered fish and that the fish were contained to prevent their escape into the wild.
  - Expanding on regulation and oversight, it is crucial to have strong national and international regulations governing the use of synthetic biology. This can include rules about containment measures, safety evaluations before any organism is released, and ongoing monitoring requirements. Regulatory agencies must be able to evaluate the potential risks and benefits of each new development, drawing on expert advice as needed.
  - Beyond governmental regulation, institutional oversight, such as through Institutional Biorisk Committees (IBCs), also plays an essential role. These committees can scrutinize research proposals, ensure researchers follow best practices, and offer guidance on biosecurity issues.
  - Public involvement in oversight can also be significant, helping to ensure that societal values and concerns are considered. This might involve public consultations or citizens' juries to debate contentious issues.
  - In all these ways, rigorous regulation and oversight can help ensure that the benefits of synthetic biology are realized while risks are minimized.

Despite these risks, it is also important to remember that synthetic biology holds immense potential for addressing many global challenges. The key is to carefully manage these risks so that we can harness the benefits of synthetic biology while minimizing potential harm.

3.5 IMPLICATIONS OF HUMAN GERMLINE EDITING

The use of synthetic biology for human germline editing, in which changes are made to the DNA of sperm, eggs, or embryos that can then be passed on to future generations, is a particularly controversial issue with many implications:

- **Medical Benefits**: Germline editing has the potential to impact the field of reproductive medicine and embryology substantially. Often, during *in vitro* fertilization (IVF) procedures, embryos are screened for genetic abnormalities before implantation, using techniques such as Pre-implantation Genetic Diagnosis (PGD). Embryos found to carry severe or lethal genetic abnormalities are typically discarded to avoid the birth of children with these conditions.

  However, with the advent of precise genome editing technologies, such as CRISPR, it may become possible to repair these genetic defects at the embryonic stage, rather than discarding the embryos. This could significantly increase the number of viable embryos available for implantation during IVF, potentially improving success rates and reducing the physical and emotional toll of multiple IVF cycles on patients.

  An example of this can be seen with conditions such as Tay-Sachs disease, which is caused by mutations in the HEXA gene. This disease leads to progressive destruction of nerve cells in the brain and spinal cord and is usually fatal in early childhood. Currently, embryos identified with Tay-Sachs disease during PGD are not implanted. However, with germline editing, it could be theoretically possible to correct the HEXA gene mutation in these embryos, preventing the disease and allowing these embryos to be implanted and develop into healthy children. This would be a paradigm shift in how we approach genetic diseases in reproductive medicine. However, the ethical and safety considerations surrounding germline editing must be carefully considered and thoroughly addressed before such procedures become a part of regular clinical practice.
- **Enhancements**: The potential application of human germline editing for human enhancements is a topic of considerable debate. While the primary aim of this technology is to prevent or cure serious genetic diseases, the capacity to make specific alterations to the human genome could theoretically be used to "enhance" certain traits or characteristics in individuals. Such enhancements could encompass a wide range of physical, cognitive, or even emotional attributes.
  - For instance, consider the potential enhancement of cognitive abilities such as intelligence. Certain genes have been associated with cognitive abilities. For example, the gene FOXO3 plays a role in intelligence and longevity. Studies have shown that individuals with certain variants of this gene tend to live longer and have better cognitive function in old age.[83] In theory, germline editing could ensure that individuals carry these "beneficial" variants, potentially enhancing their cognitive abilities and lifespan.
  - However, it is important to note that traits like intelligence are complex and likely influenced by a multitude of genes, as well as environmental factors. Moreover, the potential for unintended consequences, both medical and social, is high. Medically, we do not fully understand the interplay between different genes and editing one may have unforeseen effects due to this complex interplay. Socially, there are significant

concerns about equity (who gets access to these enhancements), consent (can you consent for your future child), and the potential for a new form of eugenics.

- This also raises significant ethical questions about the nature of human beings and the human experience. For example, if we could choose our children's traits, what does that mean for acceptance of diversity and the human condition? These are questions that society needs to grapple with as we consider the potential applications of human germline editing.

- **Unintended Consequences**: The effects of germline editing could be unpredictable and irreversible. It is possible that changes could have unintended side effects, or that they could interact with other genes in unexpected ways. These effects could then be passed on to future generations.
- **Equity Concerns**: There is a risk that germline editing could exacerbate social inequalities. If only the wealthy can access these technologies, it could lead to a genetic "elite," further dividing society along economic lines.
- **Moral and Ethical Concerns**: Many people have moral or ethical objections to germline editing. Some people believe it is wrong to interfere with human DNA in this way, while others have religious or philosophical objections.
- **Regulatory Challenges**: Regulation of germline editing is a significant challenge. Different countries have different attitudes and regulations, making it difficult to enforce these regulations in a globalized world.

In 2019, the World Health Organization established an international advisory committee to develop global standards for governance and oversight of human genome editing. The committee has stressed that germline editing for reproductive purposes is irresponsible due to significant scientific and ethical concerns.

Despite the potential risks and ethical concerns, germline editing also holds significant potential. It could potentially prevent serious genetic diseases and improve human health. However, it is crucial to have a thorough and inclusive societal debate about the implications of this technology before it is widely used.

3.6 UNINTENDED CONSEQUENCES OF SYNTHETIC BIOLOGY IN BIOMEDICAL APPLICATIONS

Synthetic biology's potential power to revolutionize the biomedical field comes with the responsibility of managing and anticipating its potential side effects. The ability to design and construct new biological components and systems, or to redesign existing ones, has profound implications.

One key risk, as mentioned, is the accidental release of engineered organisms into the environment. For instance, bacteria engineered to produce biofuels or pharmaceuticals could inadvertently be released into natural environments, compete with native species, introduce new toxins, or horizontally transfer their engineered genes to other organisms. In a worst-case scenario, such an event could lead to significant ecological disruption or even create new public health risks.

Another potential risk is the off-target effects of genetic engineering. While tools like CRISPR have significantly improved our ability to edit genomes with precision, they are not perfect and can sometimes cause mutations in unintended places. This is especially concerning in therapeutic applications, where off-target effects might lead to unexpected side effects or diseases. For example, gene therapy designed to correct a genetic disorder could inadvertently disrupt another gene and increase the risk of cancer.

Furthermore, there is also the risk of dual-use research, where research intended for beneficial purposes could be misused for harmful purposes. For example, the same techniques used to engineer a virus for therapeutic purposes could potentially create a more pathogenic or transmissible virus.

To mitigate these risks, extensive risk assessments are necessary before beginning a synthetic biology project. These should consider not only the immediate aims of the project, but also potential unintended consequences. Containment measures, such as physical barriers in laboratories or biological containment strategies (like engineering organisms to be dependent on a certain nutrient not found outside of the lab), can help prevent accidental release. Lastly, ongoing monitoring can help detect unexpected outcomes or releases as early as possible, allowing for rapid response and mitigation.

It is important to note that, while these risks are significant, they should not necessarily prevent the advancement of synthetic biology. Instead, they highlight the need for robust safety measures, ethical considerations, and regulatory frameworks that enable the responsible development of this transformative technology.

3.7 OTHER BIOSECURITY CHALLENGES IN SYNTHETIC BIOLOGY

One significant concern is the intentional misuse of synthetic biology for bioterrorism or biowarfare. As synthetic biology becomes more accessible, the risk of individuals or groups with malicious intent creating harmful or dangerous organisms increases. Ensuring robust biosecurity measures, such as strict regulation, monitoring, and responsible conduct, is vital to prevent misuse.

3.7.1 Environmental Effects in the Context of "One Health"

The "One Health" concept recognizes that the health of people is intricately connected to the health of animals and our shared environment. It is a collaborative, multisectoral, and transdisciplinary approach to achieving optimal health outcomes by recognizing the interconnection between people, animals, plants, and their shared environment.

In synthetic biology, the "One Health" approach is crucial as engineered organisms, once released into the environment, whether intentionally or accidentally, could have significant impacts not only on human health but also on animal and environmental health. Some ways this could happen include:

- **Impact on Biodiversity**: Engineered organisms could potentially outcompete natural species, leading to reduced biodiversity. Changes in one species can have knock-on effects throughout an ecosystem, potentially leading to unforeseen consequences.

- **Genetic Contamination**: Through horizontal gene transfer, engineered organisms could pass their altered genes onto natural organisms, leading to the "genetic contamination" of natural species. This could affect the health and stability of ecosystems and could potentially impact animal health if the transferred genes confer harmful traits.
- **Effect on Animal Health**: Engineered organisms might interact directly with animals in harmful ways. For example, an engineered plant might produce a substance that is harmful to certain insects, or an engineered bacterium might cause disease in animals.
- **Disruption of Ecosystem Services**: Ecosystems provide many valuable services, such as pollination, water purification, and climate regulation. The introduction of engineered organisms could disrupt these services, with potential impacts on both human and animal health.

Given these potential risks, it is crucial to consider the "One Health" approach when evaluating the safety and biosecurity of synthetic biology applications. Risk assessments should consider potential impacts on human health, animal health, and the environment, and there should be robust surveillance systems to detect and respond to any problems that arise. Efforts should be made to prevent the accidental or unauthorized release of engineered organisms, and emergency response plans should be in place in case such a release does occur. The "One Health" approach emphasizes that we need to consider the health of our entire planet to protect human health.

3.7.2 Veterinary and Wildlife Effects of Synthetic Biology

The effects of synthetic biology on veterinary and wildlife health are part of the broader "One Health" perspective, which views human, animal, and environmental health as interconnected. Some potential implications include:

- **Transgenic Animals**: Synthetic biology can create genetically engineered (transgenic) animals with altered traits. These could include livestock with increased disease resistance or enhanced productivity. While these modifications can have advantages, they also pose risks, such as potential impacts on animal welfare, changes in disease susceptibility, and effects on non-target species or ecosystems if these animals were to escape or be released.
- **Disease Control**: Synthetic biology techniques can be used to develop new methods for controlling animal diseases. For example, gene drives could theoretically reduce populations of ticks, mosquitoes, or other vectors of animal diseases. However, such approaches could also have unintended ecological impacts, and there is a risk of creating disease vectors that are more efficient or resistant to current control measures.
- **Wildlife Conservation**: Synthetic biology could potentially be used to conserve endangered species. For instance, genetic techniques could increase genetic diversity, create animals that are resistant to certain diseases, or recreate extinct species (de-extinction). However, these approaches pose significant ethical, ecological, and biosecurity challenges.

- **Biosecurity in Veterinary Settings**: Synthetic biology could lead to the creation of new pathogens or the alteration of existing ones, posing new challenges for veterinary medicine. Veterinary settings could also potentially be a source of accidental release of engineered organisms.
- **Impacts on Wildlife**: Engineered organisms released into the environment, either intentionally or accidentally, could interact with wildlife in unpredictable ways. This could impact wildlife health, behavior, and populations, and could have broader impacts on ecosystems.

3.8 CONCLUDING THOUGHTS

As we explore the boundless possibilities of synthetic biology, we are confronted with a range of biosecurity risks, from the potential misuse of synthetic organisms in bioterrorism to dilemmas surrounding DURC. The magnitude of these risks highlights the need for a proactive, comprehensive approach to biosecurity, encompassing rigorous regulatory frameworks, robust risk assessments, and a culture of responsible scientific practice. The interconnectedness of human, animal, and environmental health in synthetic biology requires a holistic "One Health" approach. It is necessary to balance the transformative power of synthetic biology with a solid commitment to biosecurity and global collaboration. The evolving nature of the field further underscores the need to adapt our strategies to address emerging risks and technologies. Balancing synthetic biology's potential benefits with the risks is pivotal in responsibly shaping its future. This balance includes considering the impacts on veterinary and wildlife health in risk assessments, regulations, and oversight. Incorporating these considerations into lab safety practices, containment strategies, response plans, and the design of synthetic biology applications is essential. Furthermore, public dialogue and ethical considerations around synthetic biology's acceptable uses and risks are crucial components of a responsible and comprehensive biosecurity strategy.

3.9 KEY TAKEAWAYS

- **Biosafety vs. Biosecurity**: Understanding the differences and intersections between biosafety and biosecurity.
- **Dual-Use Concerns**: Acknowledging the potential misuse of synthetic biology in bioterrorism and bioweapons.
- **Unintended Consequences**: Understanding the risks of accidental release of engineered organisms and their impact on ecosystems and health.
- **Ethical and Equity Issues**: Addressing ethical dilemmas in human germline editing and equity concerns in access to technologies.
- **Regulatory and Oversight Challenges**: Emphasizing the need for robust regulation and oversight in synthetic biology.
- **One Health Approach**: Recognizing the interconnectedness of human, animal, and environmental health in biosecurity considerations.
- **Data Security**: Addressing the risks of information theft and the importance of cybersecurity in synthetic biology.

3.10 THOUGHT-PROVOKING QUESTIONS

3.1 Why is it important to understand the differences between biosafety and biosecurity in synthetic biology?

3.2 How could biorisk management and biosecurity be better integrated into the fundamental concepts of synthetic biology?

3.3 How can we ensure that the advancements in synthetic biology do not outpace the development of biorisk management measures?

3.4 What role do international collaborations play in ensuring biosecurity in synthetic biology?

3.5 What measures could be implemented to better evaluate and judge the potential risks of misuse and dual-use concerns in synthetic biology?

3.6 What are the shortcomings of the measures currently in place to prevent the misuse of synthetic biology to create bioweapons?

3.7 Should academic research labs identify and list potential misuses when designing a new research project to anticipate potential risks methodically and thus develop counter-mitigation strategies and policies?

3.8 What measures should be included in an updated strategy to reduce the risk of synthetic biology being used in bioterrorism?

3.9 What parameters/elements should be included in a simulation model to assess the likelihood of a particular synthetic biology project being used to create harmful biological agents or toxins? What would be the consequences?

3.10 What parameters/elements should be included in a simulation model as the unintended consequences of releasing synthetic organisms into the environment?

3.11 What stakeholders should be part of an international committee to discuss and propose guidelines about synthetic biology, raising profound ethical questions, such as the line between therapy and enhancement or the implications of creating entirely synthetic organisms?

3.12 What measures can be taken to implement equitable access and benefit-sharing so that advancements in synthetic biology do not exacerbate existing inequities?

3.13 How can the balance between innovation and biosecurity in synthetic biology be maintained without stifling scientific progress?

3.14 What are the respective responsibilities of researchers, academic institutions, and nation-states in upholding biosecurity within the realm of synthetic biology?

3.15 How can the potential dangers of synthetic biology be effectively conveyed to the public in a manner that avoids unnecessary panic?

3.16 In what ways are projected developments in synthetic biology expected to transform the field of biosecurity in the forthcoming years?

3.17 What punitive measures should be enacted, and by which governing bodies, in response to non-compliance with international synthetic biology regulations?

4 The Role of Artificial Intelligence (AI) in Synthetic Biology and Biosecurity Implications

Artificial Intelligence (AI) is the simulation of human intelligence in machines programmed to think like humans and mimic their actions. The evolving landscape of AI technology holds immense potential to revolutionize a wide array of industries and everyday life, offering solutions that range from mundane tasks to complex problem-solving scenarios. In synthetic biology, AI emerges as a pivotal tool capable of propelling the field to new heights. This chapter delves into how AI can significantly advance synthetic biology, emphasizing its opportunities and challenges. However, AI also poses unique biosecurity concerns. Its ability to process and manipulate biological data can potentially be exploited to breach biosecurity measures. Therefore, AI is a dual-use technology serving beneficial purposes while also carrying the risk

DOI: 10.1201/9781003423171-4

of misuse for harmful objectives. This chapter will explore how AI can be responsibly harnessed in synthetic biology, focusing on the need for careful assessment and control. This chapter discusses strategies to leverage AI's strengths in advancing the field while establishing robust guardrails to prevent its deployment for malicious purposes. This balanced approach ensures that AI contributes positively to synthetic biology, enhancing its capabilities without compromising biosecurity.

4.1 AI APPLICATIONS IN SYNTHETIC BIOLOGY

AI is increasingly used in synthetic biology to enhance research capabilities, accelerate discoveries, and improve efficiency. Some of the ways AI is being applied include:

- **Gene Editing**: AI can predict the outcomes of genetic modifications using tools like CRISPR. By analyzing large datasets, AI algorithms can predict off-target effects and help optimize the design of guide RNAs, leading to more efficient and accurate gene editing.
- **Protein Design**: AI can help to predict how changes in a protein's sequence will affect its structure and function. This can speed up designing new proteins or modifying existing ones.
- **Protein Structure**: Protein structure prediction is one of the key applications of AI in synthetic biology. Proteins are the workhorses of living organisms, performing many functions that underpin life's processes. However, understanding how a protein's function is determined by its 3D structure has been one of the central challenges in biology.

  AI has revolutionized this field. Deep learning algorithms, a type of AI, have been trained to predict the 3D structure of proteins with remarkable accuracy. For instance, Google's DeepMind developed an AI system called AlphaFold that uses deep learning to predict protein structure.[84] This AI has demonstrated unparalleled precision in predicting protein structure, even besting experimental methods in some cases.

  This advancement in AI applications in synthetic biology comes with biosecurity implications. It makes synthesizing proteins and potentially harmful biological agents more accessible and less predictable, hence more difficult to monitor.
- **Genetic Circuit Design**: Designing genetic circuits, which are sequences of DNA that enable cells to perform new functions, is a complex task. AI can help to design these circuits and predict how they will behave in living cells.
- **Automated Lab Experiments**: AI can guide the automation of laboratory experiments, such as high-throughput screening or automated synthesis of genetic constructs. This can greatly speed up the research process and allow for more efficient exploration of the vast design space in synthetic biology.
- **Data Analysis**: AI has been a significant change in the analysis of genomic data, significantly accelerating the pace of discovery in synthetic biology. The volume of data produced by modern genomic technologies such as

Next-Generation Sequencing (NGS) is immense. Traditionally, analyzing these datasets required considerable time and computational resources, often taking weeks or months to complete analysis. However, AI algorithms, particularly machine learning and deep learning models, can learn from and make predictions on these data much more rapidly.

- For instance, AI models can analyze the entire genome of an organism and identify genes of interest or potential mutations associated with a disease. These models can make predictions based on the structure and sequence of these genes, as well as their interactions with other genes, accelerating the discovery process from weeks to mere days. AI models can also "learn" from data, improving their predictive capabilities over time, a feature that traditional statistical models lack.
- Moreover, AI has the potential to automate many of the analytical tasks that would typically require a programmer. For example, AI can automatically annotate a newly sequenced genome, identifying the locations of genes and other functional elements, and predicting their functions based on patterns learned from previously annotated genomes. This saves a considerable amount of time and resources, as manual annotation can be a tedious and time-consuming task.
- AI can also automate the analysis of other types of biological data. For instance, it can automate the analysis of transcriptomic data (the complete set of RNA transcripts produced by the genome), proteomic data (the complete set of proteins expressed by the genome), and metabolomic data (the complete set of small molecules or metabolites present within an organism), among others. These analyses can provide comprehensive insights into the functioning of biological systems at multiple levels, contributing to developing more effective treatments and therapies.

- **Biorisk Management and Biosecurity**: AI can predict the safety and security implications of certain research activities. For instance, it could help to predict the likelihood of accidental release of an engineered organism or to identify research activities that could potentially be misused.

While the application of AI in synthetic biology holds significant promise, it is important to consider the potential risks and challenges. These include the accuracy and reliability of AI predictions, the risk of overreliance on AI at the expense of human expertise, and the ethical and security implications of using AI in this powerful and potentially transformative field of research.

4.1.1 Application of AI Tools Used to Generate *de novo* Genes

AI has shown enormous potential in *de novo* gene synthesis, which involves designing and creating new genes from scratch. *De novo* genes can be designed to perform a specific function, such as producing a certain protein, and can be used in various applications from medical treatments to biofuels. The application of AI in *de novo* gene synthesis primarily involves two aspects:

- **Designing Gene Sequences**: AI can help design *de novo* genes by predicting the most efficient sequences to achieve a particular outcome. For example, AI algorithms can be trained on large datasets of existing gene sequences and their corresponding protein products to learn how different sequences affect protein structure and function. Using this information, they can predict the optimal gene sequence to produce a protein with desired characteristics.
- **Optimizing Gene Expression**: AI can also optimize gene expression levels. This involves designing the gene and its regulatory sequences so that it is expressed at the right level and in the right cells. AI can help with this by analyzing data on how different regulatory sequences affect gene expression and then predicting which sequences will give the desired expression level for the new gene.

These processes can involve machine learning algorithms, which can learn patterns from large amounts of data and make predictions based on these patterns, and bioinformatics tools, which can manage and analyze the large amounts of biological data involved in *de novo* gene synthesis.

Using AI for *de novo* gene synthesis can increase efficiency, reduce costs, and enable the creation of genes that might not be possible to produce using traditional methods. However, it also raises biosecurity considerations, as the same tools could be misused to create harmful genes or organisms. This highlights the need for careful regulation and oversight of AI applications in synthetic biology.

4.1.2 Employing AI Tools for Modifying Gene Sequences

Degenerating modified genes refers to the process of making calculated alterations to a gene sequence, while still retaining the same protein output. This is possible because the genetic code is redundant, meaning multiple combinations of nucleotides (codons) can code for the same amino acid. This redundancy allows for many variations of a single gene that can all code for the same protein. AI can be applied to this process in several ways:

- **Codon Optimization**: AI can help optimize codon usage for a specific host organism. For example, certain codons might be used more frequently in a particular organism, leading to higher protein production levels. AI algorithms can use large datasets of gene sequences and their corresponding protein levels to learn these patterns and predict the optimal codon usage for a given gene in a specific organism.
- **Protein Engineering**: In protein engineering, AI can assist in designing genes that code for proteins with specific desired traits. Machine learning algorithms can be trained on datasets of gene sequences and their resulting protein characteristics to learn how changes in the gene sequence affect the protein. These algorithms can then design gene sequences resulting in proteins with the desired characteristics.

- **Gene Therapies and Vaccines**: AI can design gene sequences for use in gene therapies or vaccines. For instance, AI could design a gene that codes for a specific antigen to be used in a vaccine or a gene that can correct a genetic disorder when introduced into a patient's cells.

These are just a few examples of how AI can degenerate modified genes in synthetic biology. The use of AI not only increases the efficiency and effectiveness of these processes, but it also opens up new possibilities for what can be achieved in the field. At the same time, it is important to consider the potential risks and ethical implications associated with these technologies, as well as to implement appropriate safeguards and oversight mechanisms.

4.1.3 AI Applications in Gene Screening Prior to Manufacturing

Screening gene sequences before they are synthesized is a key step for ensuring biosecurity. This can help to prevent the unintentional or unauthorized creation of harmful organisms or biological materials. AI can greatly enhance the speed and accuracy of this process. Some ways in which AI tools can be applied include:

- **Sequence Comparison and Risk Assessment**: The potential misuse of synthetic biology for malicious purposes is a significant concern. As the field advances, it becomes increasingly feasible for rogue entities or individuals to synthesize harmful biological agents such as viruses. This is where AI, particularly machine learning algorithms, can play an integral role in biosecurity.
  - AI tools can be trained to analyze genetic sequences in the orders placed with synthetic biology companies. These tools can rapidly scan and compare the ordered sequences against databases of known harmful or regulated sequences. If a match is found, the order can be flagged for further review by human experts. This process is much faster and more accurate than manual screening, especially considering the scale and complexity of genetic data.
  - Moreover, the machine learning models used for this purpose can improve over time. As they process more data, they can learn to recognize patterns associated with harmful sequences, even if they have been intentionally obfuscated or modified. This could help prevent the creation and release of genetically engineered organisms or genetic elements that could cause harm.
  - One example of such a tool is the "GeneScreener," developed by the International Gene Synthesis Consortium (IGSC). GeneScreener uses an automated algorithm to compare ordered sequences against a database of sequences of concern, flagging any matches for further review.
  - However, while AI can significantly enhance biosecurity in synthetic biology, it is also essential to consider the associated risks. For instance, AI systems can be targets of cyberattacks aimed at disrupting their

operation or manipulating their results. Therefore, robust cybersecurity measures are needed to protect these systems.

- Furthermore, effective biosecurity in synthetic biology also requires international cooperation, as harmful biological agents do not respect national boundaries. International standards and regulations are needed to ensure consistent biosecurity practices across countries and institutions. Therefore, while AI can be a powerful tool for biosecurity, it should be considered as part of a broader, multifaceted approach to biosecurity in synthetic biology.

- **Prediction of Unintended Effects**: Machine learning algorithms can also predict the potential effects of a gene sequence once it is synthesized and introduced into a living organism. For instance, they could predict potential off-target effects or the likelihood that the sequence could confer harmful traits.
- **Screening of Dual-Use Research**: Some gene sequences may have potential "dual-use" implications, meaning they could be used for beneficial and harmful purposes. AI can identify such sequences and flag them for further review.
- **Checking Against Regulatory Lists**: Some gene sequences may be regulated or prohibited by certain jurisdictions. AI can check a gene sequence against these lists quickly.
- **Automated Workflow**: AI can automate the gene screening workflow, flagging any sequences of concern, and routing them for further review while allowing low-risk sequences to be processed quickly. This can greatly increase the speed and efficiency of the screening process.

While AI tools can greatly enhance gene screening, it is important to remember that they are not infallible and should be used as part of a comprehensive biosecurity strategy. Other measures, such as physical security, regulatory compliance, staff training, and a culture of responsibility, are also crucial for ensuring biosecurity.

4.1.4 AI-Assisted Vaccine Development

The application of AI in vaccine development has been demonstrated to have tremendous potential, especially during the COVID-19 pandemic. AI has been pivotal in accelerating various stages of vaccine development, including antigen identification, vaccine design, production, and distribution. Ways how AI can aid vaccine development include:

- **Identification of Antigens**: AI can predict which parts of a pathogen's genome code for proteins that could serve as effective antigens in a vaccine. AI algorithms can analyze large genomic databases to identify these potential antigens, greatly speeding up this part of the vaccine development process.
- **Vaccine Design**: AI can make important contributions to speed up vaccine development, an important aspect of biosecurity in the face of emerging infectious diseases.

- AI can be instrumental in expediting and refining the vaccine development process. This includes optimizing the design of vaccine components such as the antigen, adjuvant, and delivery vectors. The antigen is the vaccine component that elicits an immune response, whereas the adjuvant enhances this response, and the delivery vector carries the antigen and adjuvant into the body.
- Using machine learning algorithms, AI can analyze vast amounts of biological data to identify the most suitable antigens for a particular pathogen. Furthermore, AI can also predict the safety and efficacy of different adjuvants and delivery vectors. This can involve analyzing data from preclinical and clinical studies to identify patterns associated with adverse reactions or ineffective immune responses.
- For instance, if certain adjuvants are found to be associated with allergic reactions, AI can help to identify this pattern and recommend alternatives that are less likely to cause such reactions. Similarly, AI can help to identify delivery vectors that are more efficient at introducing the antigen and adjuvant into the body, thereby improving the vaccine's effectiveness.
- An example of AI in practice was the application of AI to expedite and optimize vaccine design against anthrax. Machine learning algorithms analyzed genomic data from *Bacillus anthracis* strains to predict novel protein antigens that could elicit a protective immune response. Antibody neutralization data from animal studies provided a rich dataset that AI mined to identify correlations between antibody levels and survival rates after the challenge. This informed the selection of the optimal antigen dose and formulation.[85]
- However, while AI holds enormous potential for improving vaccine development and safety, it is important to remember that these are complex challenges that also require extensive experimental validation and rigorous clinical testing. AI is a powerful tool that can inform and guide these processes, but it is not a replacement for them.

- **Predictive Modeling**: AI can help predict how a vaccine will perform in a population, considering factors like the distribution of different age groups, pre-existing immunity, and social behavior. This can help guide vaccine distribution strategies.
- **Personalized Vaccines**: AI can also help develop personalized vaccines, especially for diseases like cancer. By analyzing a patient's unique genetic information and the genetic makeup of their tumor, AI can aid in designing a personalized vaccine to elicit a robust immune response against the tumor.
- **Production Optimization**: AI can help optimize the production of vaccines, predicting the most efficient processes and ensuring quality control.
- **Vaccine Distribution**: AI can optimize vaccine distribution considering factors such as population density, disease prevalence, and infrastructure to ensure vaccines reach those in the greatest need.

However, the use of AI in vaccine development also comes with certain challenges, including the need for high-quality data, issues with algorithmic bias, and the need for transparency in how AI algorithms make their predictions. Furthermore, ethical considerations around using personal genetic data for vaccine development must be addressed.

4.1.5 AI-Trained Library Screenings

Screening libraries of various biological or chemical entities is a common practice in biotechnology research. However, the vastness of these libraries, which can contain millions of unique entities, poses a challenge for manual screening. AI can help in this regard by significantly improving the speed, cost-effectiveness, and outcomes of library screening. Applications of AI in screening several types of libraries include:

- **Chemical Libraries**: AI can predict which chemicals in a library are most likely to have a desired effect, such as inhibiting a specific protein or serving as a potential drug candidate. AI algorithms can learn from data on the structures and activities of known chemicals to make these predictions.
- **Antibody Libraries**: In monoclonal and single-chain antibody libraries, AI can help predict which antibodies are most likely to bind a specific antigen with high affinity and specificity. It can also design new antibodies or optimize existing ones.
- **Polymer Libraries**: AI can help predict the properties of polymers based on their chemical structure, such as their mechanical properties, degradation rate, or biocompatibility. This can be useful in areas like materials science, drug delivery, and tissue engineering.
- **Virus-Like Particle (VLP) Libraries**: AI can design and optimize VLPs for use in vaccines or other therapies. This can involve predicting how changes in the VLP structure will affect its properties or optimizing the design of VLPs to elicit a specific immune response.
- **Directed Evolution of Specific Targets**: AI can guide the process of directed evolution, which involves iterative rounds of mutation and selection to optimize a protein or organism for a specific function. AI can predict which mutations will be most likely to improve function, greatly speeding up this process.
- **Bacterial Libraries**: In bacterial libraries, AI can predict which strains are most likely to have desired properties, such as high production of a specific metabolite or resistance to a stressor. It can also design new strains or optimize existing ones.
- **Environmental Sample Libraries**: AI can help analyze libraries of environmental samples, such as soil or water samples, to predict which samples are most likely to contain a desired organism or gene. This can be useful in areas like bioprospecting, where researchers search for novel organisms or genes with potential biotechnological applications.

These applications of AI can greatly increase the speed and efficiency of library screening and open new possibilities for discovery. However, they also require large amounts of high-quality data, and the predictions made by AI algorithms must be validated experimentally. As such, the use of AI in library screening should be seen as a tool to guide and augment human research, rather than a replacement for human expertise.

4.1.6 Benefits of AI Modeling in Drug Screening

AI has made significant contributions to drug screening processes, bringing several benefits that can speed up and enhance drug discovery:

- **Efficiency**: Using AI for drug screening can greatly speed up the process. Traditional methods of drug screening, which involve testing each compound individually, can be time-consuming and costly. AI can analyze large databases of compounds and predict their likely effects, reducing the need for extensive lab testing.
- **Precision**: AI can help to increase the precision of drug screening by considering a wide range of factors that may influence a drug's effects. This can include everything from the structure of the drug molecule to the genetic characteristics of the target organism.
- **Predictive Capabilities**: With AI, researchers can predict the effects of drugs that have not yet been synthesized. This can help to guide the drug design process and avoid the synthesis of compounds that are unlikely to be effective.
- **Reduced Costs**: By streamlining the drug screening process and reducing the need for extensive lab testing, AI can help to reduce the costs associated with drug discovery.
- **Personalized Medicine**: AI algorithms can be trained to predict how different individuals will respond to a given drug based on their genetic profile, potentially paving the way for more personalized approaches to medicine.
- **Repurposing Drugs**: AI can help identify new uses for existing drugs. By analyzing the properties of a drug and comparing them with the requirements for treating a certain condition, AI can suggest potential new uses for drugs that have already been approved.

The use of AI comes with challenges, such as the need for high-quality data, potential bias in AI predictions, and the need to validate AI predictions with lab testing. Despite these challenges, the potential benefits of AI are substantial. For instance, the data generated by AI for protein structure prediction, drug discovery, and gene library screening is proprietary, sensitive, and can be a target for cyberattacks. While this technology holds significant promise, it must be complemented by robust biosecurity measures and ethical guidelines to ensure that it is used responsibly.

4.2 USING AI TOOLS TO MANAGE AND MINE LARGE DATASETS

The increasing integration of AI in synthetic biology has led to the generation of massive amounts of data, ranging from genomic sequences to protein structures and metabolic pathways. This large-scale data is a gold mine for new insights, but managing and making sense of this data is a significant challenge. AI and machine learning tools can provide immense help in this regard:

- **Data Management**: AI can help manage vast databases by organizing, labeling, and making connections between data points. This can greatly speed up finding relevant data, which can be a bottleneck when dealing with large datasets.
- **Pattern Recognition**: AI algorithms, especially those using machine learning, excel at recognizing patterns in large datasets that would be impossible for a human to discern. This can lead to new insights and predictions, such as identifying new gene functions or predicting how a protein will fold based on its sequence.
- **Predictive Modeling**: Machine learning can create predictive models based on large datasets. For instance, algorithms can be trained on data about the structure and function of known proteins to predict the function of a newly sequenced protein.
- **Automated Analysis**: AI can automate repetitive analysis tasks, significantly speeding up data processing. For instance, AI can annotate new genomic sequences automatically or identify key features in microscopy images.
- **Integrative Analysis**: AI can integrate and analyze data from multiple sources to provide a more comprehensive understanding of biological systems. For example, genomic, transcriptomic, and proteomic data can be integrated to provide a holistic view of a cell's function.
- **Decision-Making Support**: AI can help researchers make decisions based on large amounts of data. For example, it could help decide which experiments to run next or which potential drug candidate to synthesize and test.
- **Biosecurity**: AI can analyze large datasets to identify potential risks, such as sequences that could be used to engineer harmful organisms.

AI offers powerful tools for managing and mining large datasets, but it is important to consider potential challenges, such as biases in how AI models interpret data. Furthermore, it is crucial to validate AI-derived insights with experimental data to ensure their accuracy and relevance.

4.3 USING AI TOOLS FOR PUBLIC HEALTH SURVEILLANCE

AI tools have been increasingly applied for public health surveillance and have demonstrated considerable potential in improving the detection, monitoring, and response to various health threats. The way how AI tools can be utilized in public health surveillance include:

- **Disease Detection and Forecasting**: AI algorithms can analyze various data sources, such as social media posts, news reports, and health records, to detect potential disease outbreaks in real time. AI can also forecast the spread of diseases based on factors like population movement, environmental conditions, and health infrastructure.

- **Genomic Surveillance**: AI can aid in analyzing and interpreting vast amounts of genomic data generated from pathogen sequencing. This can help monitor the emergence of new strains or variants and assess their potential impacts on public health.
- **Monitoring Health Behaviors**: AI tools can analyze data from sources like wearable devices or online search queries to monitor health behaviors and trends at the population level, such as physical activity levels, diet, or mental health status.
- **Healthcare System Monitoring**: AI can analyze data from healthcare systems to detect issues such as hospital overcrowding, supply shortages, or delays in care, enabling timely responses to these problems.
- **Biosecurity Threat Detection**: AI can analyze data from various sources to detect potential biosecurity threats, such as unusual disease outbreaks that could indicate the use of a biological weapon or the accidental release of an engineered organism.
- **Public Health Policy Evaluation**: AI can analyze data on the impacts of public health policies, helping to evaluate their effectiveness and guide future policy decisions.

While AI holds great potential for improving public health surveillance, it is important to consider the associated challenges. These include ensuring data privacy and security, addressing issues of data bias and quality, and ensuring that AI tools are used in a way that complements, rather than replaces, human expertise. Public health decisions based on AI should always be guided by ethical considerations and an understanding of the social and cultural context.

4.4 INCORPORATING AI IN COMPREHENSIVE "ONE HEALTH" SOLUTIONS

The "One Health" approach recognizes the interconnectedness of human, animal, and environmental health. With the advancement of AI, the "One Health" approach can be bolstered and contribute to solving global health problems more effectively. Some of the ways AI can be integrated into "One Health" solutions include:

- **Disease Surveillance and Prediction**: AI can analyze data from diverse sources, including human health records, wildlife disease monitoring, climate data, social media, send requests from first responders/ambulance support, etc., to predict and track disease outbreaks that can jump between animals, humans, and environmental triggers, such as zoonotic diseases.
- **Genomic Analysis**: AI can help analyze and interpret large genomic datasets from various species (human, animal, plant, and microorganisms), leading to insights about disease resistance, vulnerability, and transmission patterns across species.
- **Environmental Health**: AI can analyze vast environmental data, including air quality, climate change parameters, water quality, deforestation rates, etc., to understand and predict their impacts on human and animal health.

- **Food Safety**: AI can monitor and predict potential threats to food safety, such as predicting the presence of toxins or harmful microbes in crops or livestock based on environmental and health data.
- **Drug and Vaccine Development**: AI can speed up drug and vaccine development for diseases impacting human and animal health. AI models can predict promising drug candidates or vaccine antigenic components, saving considerable time and resources.
- **Policymaking**: AI can aid in policy decision-making by forecasting the impact of policy changes on human, animal, and environmental health.
- **Healthcare Delivery**: AI can optimize the delivery of healthcare services to remote or underserved populations, including veterinary services.

By considering all aspects of the ecosystem, the "One Health" approach aims to achieve optimal health outcomes. AI, with its ability to analyze and derive insights from vast and diverse data, can enhance this holistic approach. However, potential biases in AI models and the need for multisectoral collaborations need to be addressed to fully leverage AI's potential in "One Health" solutions.

4.5 VULNERABILITIES OF USING AI TECHNOLOGY

While AI holds tremendous potential for advancing synthetic biology and improving biosecurity, it also introduces several vulnerabilities and challenges that need to be addressed. As an example of this urgent issue being recognized by governments, in 2023, the White House issued an executive order[86] to address AI's integration across various fields, including synthetic biology. Therefore, the biorisk management community must quickly adapt and consider AI as another factor when conducting a biosecurity risk assessment. Some key considerations include:

- **Data Privacy and Security**: The amalgamation of vast datasets is one of the key drivers behind the burgeoning power of AI in synthetic biology and biosecurity. Through access to larger and more diverse datasets, AI algorithms can uncover richer patterns and make more accurate predictions, thus potentially catalyzing major breakthroughs. For instance, datasets from multinational research groups can provide diverse data points, thus enhancing the accuracy of AI models. Similarly, pooling global health records can give unprecedented insights into disease patterns and treatment outcomes, which can then be leveraged to create personalized treatment strategies or to respond to emerging health crises rapidly.
    - However, the path to these potential breakthroughs is littered with formidable challenges. The exchange and pooling of data across borders can trigger suspicions among nations, often related to concerns over national security, commercial competitiveness, or the misuse of citizens' health data. Furthermore, data quality is a significant concern, as data from diverse sources may vary in reliability and consistency.
    - A prime example of these challenges was seen during the COVID-19 pandemic. Global collaboration and data sharing were vital for tracking

the virus's spread and developing treatments and vaccines. However, differences in data collection methods, transparency issues, and geopolitical tensions sometimes hindered these efforts.[87–89]

- Addressing these challenges will require the creation of robust guidelines, regulations, and international agreements. Such measures must aim to promote transparency, consistency, and security in data-sharing practices, while also ensuring that all countries have equitable access to the benefits of AI.
- Regulations should define what data can be shared, who can access it, and how it should be used, while also including strong protections for privacy and confidentiality. They should also provide mechanisms for addressing discrepancies in data quality and for verifying the accuracy of data.
- International agreements, negotiated through diplomatic channels or global institutions like the World Health Organization, can provide a framework for global data sharing. These agreements should be based on principles of mutual benefit, respect for national sovereignty, and the global common good.
- In addition to these regulatory and diplomatic efforts, technological solutions like differential privacy, federated learning, and blockchain can also play a role in promoting secure and privacy-preserving data sharing. For instance, differential privacy allows for extraction of useful insights from datasets while keeping the data of individual participants anonymous. Federated learning enables AI models to learn from decentralized data sources, reducing the need to centralize sensitive data. Blockchain can provide a secure and transparent platform for data sharing, with an immutable record of all transactions.

- **Data Quality and Bias**: AI models are only as good as the data they are trained on. If the training data is of poor quality or contains biases, this can lead to inaccurate or biased predictions. This is a significant concern in healthcare applications, where biases can have serious consequences.
- **Transparency and Explainability**: AI models, especially those using complex machine learning techniques, can be "black boxes," meaning it is difficult to understand how they make their predictions. This lack of transparency can make it difficult to trust AI predictions, especially in high-stakes areas like healthcare or biosecurity.
- **Reliability and Validation**: AI models need to be rigorously validated to ensure that their predictions are reliable. This is particularly important in synthetic biology, where incorrect predictions could lead to harmful outcomes.
- **Misuse of Technology**: As with any powerful technology, there is a risk that AI could be misused. For instance, AI could potentially be used to design harmful biological agents or to circumvent biosecurity measures.

Addressing these vulnerabilities requires a multifaceted approach, including robust data security measures, thorough validation of AI models, transparency and accountability in AI development, and strong ethical guidelines and regulations. It also requires ongoing dialogue between researchers, policymakers, and the public to navigate the ethical and societal implications of this rapidly evolving field.

4.5.1 Risks of Data and Intellectual Property Theft Using AI Technology

AI and synthetic biology are rapidly advancing fields that offer significant scientific, technological, and commercial opportunities. They also, however, present attractive targets for data and intellectual property (IP) theft. In this vein, there is a growing concern that AI tools, in the wrong hands, could become powerful enablers for cyberattacks, facilitating the unauthorized access and theft of IP and data. Here is a list of vulnerabilities:

- **Data**: AI systems often rely on large datasets, which may contain sensitive or proprietary information. Unauthorized access to these data can lead to the theft of valuable insights, breach of privacy, or misuse of the data. For instance, genomic data could reveal information about an individual's health or genetic traits, or the characteristics of a novel organism or bioengineered product.
- **IP**: AI algorithms, biological designs, and techniques developed in synthetic biology often represent significant IP. These can be targets for theft, potentially leading to significant financial loss or competitive disadvantage. For example, the design of a novel bioengineered organism or the code of an AI system used for gene editing could be valuable IP.
- **AI**: AI algorithms can create sophisticated hacking tools capable of infiltrating even highly secure systems. These tools can potentially automate and scale attacks, making them more difficult to prevent and detect. They can use techniques such as machine learning to adapt and evolve, bypassing security measures and exploiting new vulnerabilities as they emerge. For instance, AI can be used to create advanced phishing techniques where fraudulent emails can be personalized and sent in large volumes, effectively deceiving even the wariest of recipients. Deep learning algorithms can generate realistic deepfake audios or videos to manipulate people into sharing sensitive information or to create discord and misinformation.

  AI can also perform rapid, automated vulnerability scanning across vast networks, identifying weak points far more quickly than human hackers. Once an entry point has been identified, AI can also automate the exploitation of these vulnerabilities, gaining unauthorized access to systems and extracting sensitive data or IP. Regarding IP theft, AI could be used to steal proprietary algorithms or datasets, industrial designs, business strategies, or other valuable corporate information. The impacts of such theft could be enormous, leading to significant financial losses and undermining competitive advantages.

  All these potential risks highlight the importance of investing in robust cybersecurity measures to keep pace with these rapidly evolving threats. It is essential to integrate AI into our defense mechanisms as well, using AI-driven threat detection and response tools that can match the sophistication of AI-enabled cyber threats.

- **Biosecurity Risks**: Data and IP theft in the realm of AI and synthetic biology is not just a potential business risk; it is a matter of national security. Infiltration of databases and unauthorized access to AI models or synthetic biology blueprints can lead to the leak of valuable IP. This can have serious implications ranging from losing a competitive edge to endangering public health and safety.
  - If a country loses its proprietary data to another, this could result in a substantial setback for the originating country, while unfairly providing the infiltrating country with an undeserved technological leap forward. More importantly, data theft could contribute to a national security risk if the stolen information is used inappropriately or maliciously. It could lead to the engineering of harmful biological agents or the design of biotechnologies that could be exploited for biological warfare or bioterrorism.
  - The advancement of AI and synthetic biology is accelerating the rate at which we can manipulate biological systems down to the genetic level. While this presents promising prospects for medicine and biotechnology, it also means that in the wrong hands, such capabilities could be misused. For instance, a nefarious actor with access to synthetic biology techniques could potentially design harmful pathogens or, with access to certain AI models, could potentially predict and exploit vulnerabilities in biosecurity measures.
  - Moreover, this risk extends beyond state actors. In today's digital world, information theft is not just a concern between nations; it is also a considerable risk involving non-state actors, such as hackers, organized crime groups, or terrorist organizations. These entities might be motivated by financial gain, ideological beliefs, or a desire to sow chaos and fear.
  - International cooperation is key in addressing this issue. Countries should work together to establish global norms and regulations regarding data security and IP protection in AI and synthetic biology. Such cooperation could involve sharing information about threats and best practices, collaborating on investigations, and jointly responding to incidents. It is also essential to ensure that such agreements and practices are fair and equitable, acknowledging countries' different resources and capacities. Policies should include mitigation measures and consequences for countries that refuse to collaborate to prevent global harm.

Addressing the risks associated with data and IP theft in AI and synthetic biology necessitates implementing of strong data security measures. Encryption of sensitive data, robust access controls, and use of secure networks are vital first steps in this process.

In addition to these safeguards, implementing redundancy in data storage or backing up essential data in different physical locations can add an extra layer of security. This is particularly crucial in mitigating the risks of data loss due to unforeseen events like natural disasters, hardware failures, or cyberattacks. Redundant data storage can also enable rapid recovery of data, which is crucial for maintaining business continuity and preventing significant disruption of services.

Companies and research institutions should not only focus on technical measures but also invest in creating and enforcing clear IP policies. These may include obtaining patents, using trade secrets, and setting up non-disclosure agreements. This legal framework can protect proprietary technologies and research findings from being used without permission.

Moreover, human error is often a significant factor in data breaches. To minimize this risk, providing education and training on data security and IP protection can be integral to a comprehensive security strategy. Training programs can keep all staff members – from frontline researchers to top management – aware of the potential risks, familiar with the security protocols, and equipped with the knowledge to identify and prevent potential security breaches.

Finally, it is important to recognize that data and IP security measures need to be continually updated and adapted. With the rapid advancement of technology, new vulnerabilities can emerge and cyber threats can evolve. Regular reviews of security policies and practices, in addition to staying abreast of the latest developments in cybersecurity, can help maintain a robust defense against data and IP theft.

Furthermore, monitoring and response strategies should be in place to identify and respond to incidents of data or IP theft. This might involve cyber threat monitoring, regular security audits, and incident response plans. Lastly, international cooperation and legal enforcement can also be important for addressing these risks at a global level.

4.6 UNINTENDED CONSEQUENCES OF AI APPLICATIONS IN SYNTHETIC BIOLOGY

While using AI in synthetic biology presents enormous potential for advancements, it also poses risks of unintended consequences. These unintended consequences could have significant implications for both individuals and society. Some examples include:

- **Misinterpretation of AI Predictions**: AI algorithms in synthetic biology usually work with complex data and generate predictive models based on this data. However, there is a risk that these predictions could be misinterpreted or overly relied upon without proper validation, leading to incorrect conclusions or inappropriate actions.
- **Algorithmic Bias**: If the data used to train AI algorithms contain biases, the predictions made by these algorithms could also be biased. In synthetic biology, this could lead to biased predictions about disease risks, drug responses, or other biological phenomena.
- **Unanticipated Effects of Gene Editing**: AI is increasingly used to guide gene editing. While this can increase efficiency, there is also a risk of unanticipated negative effects. For instance, off-target effects could lead to unintended modifications in the genome, which could have harmful consequences.
- **Biosecurity Risks**: AI could potentially design harmful biological agents or circumvent biosecurity measures. This could present significant biosecurity risks.

- **Ethical Implications**: Using AI in synthetic biology raises several ethical questions. For instance, who is responsible when an AI-guided experiment produces harmful outcomes? How should the benefits and risks of these technologies be distributed? How should privacy be protected when AI analyzes personal genetic data?

Preventing these unintended consequences requires a multifaceted approach, including robust scientific validation of AI predictions, addressing biases in training data, comprehensive biosecurity measures, careful consideration of ethical issues, and ongoing dialogue and regulation to navigate the societal implications.

4.7 BIOSECURITY GAPS IN AI TECHNOLOGY

AI and synthetic biology are rapidly evolving fields, and with their development, several biosecurity gaps have emerged that present significant challenges:

- **Gaps in Policy and Guidance**: There are currently no universally accepted regulations specifically governing the use of AI in synthetic biology. Policies and guidance are still in their early stages and may not fully address all potential issues. For example, policies may not sufficiently cover the use of AI for predicting gene function or guiding gene editing. The lack of clear guidelines can create uncertainty and potential security risks.
- **Technology Outpacing Oversight and Scientific Consensus**: The pace of technological advancement in AI and synthetic biology often outstrips the development of oversight mechanisms and the establishment of scientific consensus. This can make it challenging for regulatory bodies and the broader scientific community to keep up with the latest developments and understand their potential implications. This gap can result in inadequate oversight and the potential for misuse of technology.
- **Uneven Adoption of AI in the Scientific Community**: The use of AI in synthetic biology is not uniformly distributed across the scientific community. Some researchers and institutions may have extensive resources and expertise in AI, while others may have little to no access to these tools. This uneven adoption can lead to disparities in research capabilities and outputs, and it may also impact the robustness of biosecurity measures across different institutions.

To address these gaps, efforts are needed on multiple fronts. Policymakers, researchers, and biosecurity experts need to collaborate to develop comprehensive and adaptable policies and guidelines. Continued efforts are needed to build consensus on best practices and to develop robust oversight mechanisms that can keep pace with technological advancements. Additionally, initiatives to democratize access to AI tools and training can help ensure uniform adoption across the scientific community and strengthen overall biosecurity. Finally, fostering a strong culture of responsibility and ethics in AI and synthetic biology can help mitigate biosecurity risks.

4.8 BIOSECURITY RISK ASSESSMENT GUIDELINES FOR AI USE IN SYNTHETIC BIOLOGY

This section explores the safe application of AI in synthetic biology, emphasizing the importance of thorough biosecurity risk assessment and management. This section introduces a systematic guide for identifying and evaluating the biosecurity risks associated with integrating AI into this field. This robust methodology, supported by practical tools (referenced in Tables 4.1–4.5), aids biorisk management professionals in understanding and mitigating potential risks while fostering the progress of AI in synthetic biology.[90]

This framework provides a nuanced view of the risk landscape, taking into account factors such as the level of automation, AI technology maturity, and AI model types. It offers a structured approach to conducting comprehensive biosecurity risk assessments, focusing on understanding specific AI technologies, assessing risks and vulnerabilities, and implementing effective mitigation strategies for responsible and secure AI use in synthetic biology.

Additionally, it addresses significant biosecurity concerns that new AI may introduce in synthetic biology, such as the potential misuse of AI in creating harmful biological agents or data security breaches. This section presents innovative tools and methodologies for comprehensive biosecurity risk assessments in AI-driven synthetic biology, enabling professionals to critically evaluate and develop effective mitigation strategies. This guidance paves the way for more informed and secure AI applications in the field.

4.8.1 Step 1: AI Application and Its Contextual Risks

The first step in performing an AI biorisk assessment is to identify the specific AI applications to be used and the specific experiments in synthetic biology. Use Table 4.1 to guide this identification process. The same table can then be used to define or categorize the relevant risk in synthetic biology applications.

4.8.2 Step 2: Threats and Vulnerabilities

The second step requires assessing the various vulnerabilities and threats of AI technology. Table 4.2 has been developed to identify applicable vulnerabilities or threats of the AI system being assessed.

4.8.3 Step 3: Maturity and Automation Level

The third step entails evaluating the AI system's maturity and automation level. Table 4.3 was developed to assess the maturity of the AI system (emerging, current, obsolete, etc.), and Table 4.4 to assess its level of automation. Table 4.4 breaks down AI systems into seven degrees of automation, from no automation to full autonomy, and provides insights into the degree of human control, system control, and associated risk level for each.[110,111] These two tables when combined help to understand the potential risks and the degree of human oversight required.

TABLE 4.1
Summary of AI Applications in Synthetic Biology

Application	Explanation	Threats/Vulnerabilities	Risk
Gene editing[91]	AI can help to predict the outcomes of genetic modifications using tools like CRISPR. By analyzing large datasets, AI algorithms can predict off-target effects and help optimize the design of guide RNAs, leading to more efficient and accurate gene editing	• Risk of dual use • Ethical and safety concerns • Lack of oversight and quality control • Risk of unintended consequences • Multiple dependencies on the hardware and the software (makes the system fragile)	High
De novo gene design[92]	AI has shown enormous potential in *de novo* gene synthesis, which involves designing and creating new genes from scratch. *De novo* genes can be designed to perform a specific function, such as producing a certain protein, and can be used in various applications, from medical treatments to biofuels	• High level of complexity and uncertainty (this is truly very difficult to achieve) • Risk of dual use • High potential for unintended consequences • Potential for misuse • Oversight challenges • Lack of technological maturity and expertise	Very high (however difficult to achieve)
Gene sequence modification[91]	Degenerating modified genes refers to the process of making calculated alterations to a gene sequence while still retaining the same protein output. This is possible because the genetic code is redundant, meaning multiple combinations of nucleotides (codons) can code for the same amino acid. This redundancy allows for many variations of a single gene that can all code for the same protein	Although this is technically very simple to catch there is still vulnerability: • Potential for misuse • Off-target effects and unintended consequences • Technical complexity	High
Gene screening prior to manufacturing[93]	AI can help screen gene sequences before they are synthesized. This can help to prevent the unintentional or unauthorized creation of harmful organisms or biological materials	• Controlled and targeted process • Purpose of optimization • Potential for unintended effects • Potential for misuse • Oversight challenges • Technical expertise and safeguards impact on manufacturing and end-use	Moderate

(*Continued*)

TABLES 4.1 (*Continued*)
Summary of AI Applications in Synthetic Biology

Application	Explanation	Threats/Vulnerabilities	Risk
Protein design[94]	AI can help to predict how changes in a protein's sequence will affect its structure and function. This can speed up designing new proteins or modifying existing ones	• Risk of dual use • Precision and efficiency • Complexity of protein design • Potential for unintended consequences • Biosecurity and ethical considerations • Dependency on AI predictions • Oversight challenges	Moderate
Protein structure[84]	Protein structure prediction is one of the key applications of AI in synthetic biology. Understanding how a protein's function is determined by its 3D structure is one of the central challenges in biology. This is because the number of configurations a protein can fold into is astronomically large, making it practically impossible to predict a protein's structure based solely on its sequence of amino acids using conventional methods. AI can help accurately predict protein folding based on sequences	There is Alpha Fold[84] and API[95] open-source AI, technically simple to do. However, there are still vulnerabilities: • Advances in AI-powered prediction • Importance of accurate prediction • Limitations of predictive models • Potential for misinterpretation or misuse • Dependence on data quality • Oversight challenges	Moderate
Vaccine development[96]	AI has been pivotal in accelerating various stages of vaccine development, including antigen identification, vaccine design, production, and distribution	• Enhanced research efficiency • Data-driven insights • Improved accuracy and predictive power • Supportive role in a regulated environment • Oversight challenges • Risk of dependence on AI predictions • Public health consequences	Low

(*Continued*)

TABLES 4.1 (*Continued*)
Summary of AI Applications in Synthetic Biology

Application	Explanation	Threats/Vulnerabilities	Risk
Genetic circuit design[97]	Designing genetic circuits, which are sequences of DNA that enable cells to perform new functions, is a complex task. AI can help to design these circuits and predict how they will behave in living cells	This is technically difficult to do: • Complexity of genetic circuits • Risk of dual use • Potential for unintended consequences • Biosecurity concerns • Oversight challenges • Technical expertise and precision required • Dependence on modeling and predictive tools • Advancements in safety and control mechanisms	Moderate
Data analysis[98]	AI has been a significant change in the analysis of genomic data, accelerating the pace of discovery in synthetic biology. The volume of data produced by modern genomic technologies such as Next-Generation Sequencing (NGS) is immense. AI algorithms, particularly machine learning and deep learning models, can learn from and make predictions on these rapidly, making the analysis possible on the scale of hours to days	• Non-invasive nature of analysis • Highly regulated data handling • Advances in computational techniques • Supportive role in research • Oversight challenges • Potential for data privacy concerns	Low
Library screening[99]	AI can help screen libraries of various biological or chemical entities, significantly improving the speed, cost-effectiveness, and outcomes of library screening	• Controlled and targeted screening process • Standardized protocols and procedures • No direct genetic manipulation • Use in drug discovery and development • High-throughput and automated systems • Ethical and regulatory compliance	Low

(*Continued*)

TABLES 4.1 (*Continued*)
Summary of AI Applications in Synthetic Biology

Application	Explanation	Threats/Vulnerabilities	Risk
Drug screening[100]	AI can greatly speed up the process of drug screening. AI can analyze large databases of compounds and predict their likely effects, reducing the need for extensive lab testing	• Chemical and biological interactions • Risk of dual use • High-throughput and automated systems • Potential for misinterpretation • Regulatory compliance • Biosecurity considerations • Safety protocols in laboratory settings	Moderate
Automated lab experiments[101]	AI can guide the automation of laboratory experiments. This can greatly speed up the research process and make it more efficient	• Complexity and variability of experiments • Risk of dual use • Overreliance on technology • Potential for equipment failure or malfunction • Potential for misuse • Biosecurity and containment risks • Lack of ethical and regulatory standards • Data integrity and reproducibility	High
Biorisk management and biosecurity	AI can predict the safety and security implications of certain research activities. For instance, it could help predict the likelihood of an engineered organism's accidental release or identify research activities that could be misused	• Sensitive information handling • Dependence on AI accuracy and reliability • Complex ethical implications • Risk of dual use • Security of AI systems • Requirement for expert oversight • Regulatory and compliance challenges	High

These are the main synthetic biology applications where AI plays an important role, while raising biosecurity concerns. The risk levels were determined by the author in consultation with experts in the fields of AI and biosecurity. They are meant to be starting points that can be adapted to different circumstances. The table also details their specific functions, explanations, and associated threats/vulnerabilities.

TABLE 4.2
Summary of the Vulnerabilities and Challenges of AI Applications in Synthetic Biology

Vulnerability	Explanation
1. Data privacy and security	Through access to larger and more diverse datasets, AI algorithms can uncover richer patterns and make more accurate predictions. Using datasets from multinational research groups can provide diverse data points that enhance the accuracy of AI models. Pooling global health records can give insights into disease patterns and treatment outcomes, which can then be leveraged to create personalized treatment strategies or rapidly respond to emerging health crises. Technological solutions like differential privacy,[102] federated learning,[103] and blockchain[104] can also play a role in promoting secure and privacy-preserving data sharing. For instance, differential privacy allows for extracting useful insights from datasets while keeping the data of individual participants anonymous. Federated learning enables AI models to learn from decentralized data sources, reducing the need to centralize sensitive data. Blockchain can provide a secure and transparent platform for data sharing, with an immutable record of all transactions
2. Data quality and bias	AI models are only as good as the data they are trained on. If the training data is of poor quality or contains biases, this can lead to inaccurate or biased predictions. • Data quality is a significant concern, as data from diverse sources may vary in reliability and consistency. • During the COVID-19 pandemic, global collaboration and data sharing were vital for tracking the virus's spread and developing treatments and vaccines. However, differences in data collection methods, transparency issues, and geopolitical tensions sometimes hindered these efforts[87–89]
3. Transparency and explainability	AI models, especially those using complex machine learning techniques, can be "black boxes," making it difficult to understand how they make their predictions. This lack of transparency can make it difficult to trust AI predictions, especially in high-stakes areas like healthcare or biosecurity.[105,106] It is important to note that more recently developed AI is more transparent, and this area is rapidly evolving
4. Reliability and validation	AI models need to be rigorously validated to ensure that their predictions are reliable. This is particularly important in synthetic biology, where incorrect predictions could lead to harmful outcomes[107]
5. Data and IP theft	AI and synthetic biology are rapidly advancing fields that offer significant scientific, technological, and commercial opportunities. They present attractive targets for data and intellectual property (IP) theft, and there is a growing concern that AI tools, in the wrong hands, could become powerful enablers for cyberattacks, facilitating the unauthorized access and theft of IP and data[108,109]

Among the various key AI vulnerabilities, data privacy and security issues can pose risks of data breaches or misuse due to the extensive data requirements of AI systems. The risk assessor should note that not every listed vulnerability applies universally, and that this table should be used to assess specific vulnerabilities in the specific AI system being assessed.

TABLE 4.3
Maturity Levels of AI Systems

Maturity Level of AI Technology	Description	Relative Risk Level	Threats/Vulnerabilities
Emerging	AI systems are in the early phases of capability building, characterized by basic functionalities, limited scope, and a primary focus on exploration and learning. Emerging AI often involves rudimentary algorithms that can perform simple tasks or analyses, but lack the advanced features, depth, and sophistication of more mature AI systems	High	• Limited predictability and control • Lack of advanced safety and ethical protocols • Potential for misuse or misinterpretation • Need for significant human oversight • Rapidly evolving technology
Limited	The technology is operational for the implementation of a limited number of applications	Moderate	• Defined but narrow capabilities • Improved safety and ethical standards • Requirement for human oversight • Potential for misinterpretation or overreliance • Incremental improvements and learning
Strategic	AI capabilities are more defined and focused, capable of handling specific tasks with a reasonable degree of efficiency. However, these systems still exhibit complexity, scope, and adaptability constraints. AI functionalities at this stage are often restricted to narrow domains or particular types of tasks, with limited ability to generalize or adapt to new or unforeseen challenges	Moderate	• Advanced capabilities with a specific focus • Better integration and autonomy • Enhanced ethical and safety protocols • Potential for overreliance • Need for ongoing monitoring and evaluation
Preferred	Highly advanced stage in AI development, characterized by AI systems that are sophisticated in their capabilities and broadly recognized as reliable and effective solutions in their respective domains	Low	• Advanced autonomy with robust safeguards • High reliability and proven track record • Deep integration and understanding • Enhanced learning and adaptation capabilities • Comprehensive compliance with ethical and regulatory standards • User trust and dependency

(*Continued*)

TABLE 4.3 (*Continued*)
Maturity Levels of AI Systems

Maturity Level of AI Technology	Description	Relative Risk Level	Threats/Vulnerabilities
Current	The forefront of AI development, embodying the most advanced, state-of-the-art capabilities available in the field	Low	• Advanced and adaptive safety protocols • High-level autonomy with responsible oversight • Proven reliability and effectiveness • Sophisticated real-time learning and adaptation • Compliance with regulatory standards • Widespread trust and acceptance
Obsolete	AI systems have become outdated in terms of technology, functionality, and relevance	High	• Outdated technology and limited capabilities • Security vulnerabilities • Incompatibility with current standards • Lack of support and updates • Potential for misuse • Reduced user trust and reliance

These levels are ranked from the lowest maturity level ("emerging") to the overly mature one ("obsolete"), along with the associated characteristics and vulnerabilities. This table helps risk assessors understand how the developmental stage of an AI system influences its risk profile.

4.8.4 Step 4: Determine the Consequences

The fourth step requires determining the consequences and risk levels. Table 4.5 was developed to determine the potential consequences if certain risks materialize. The table also suggests mitigation strategies for each risk level.[113] Risk assessors should assign a risk level (low, moderate, and high) to each potential consequence based on its severity and probability of occurrence. They should use this table to systematically assess risks and implement appropriate mitigation strategies in their AI applications.

4.8.5 Steps 5–7: Mitigation, Monitor, and Review

The fifth step is to develop and implement various mitigation strategies. For each identified risk, the risk assessor should develop strategies to mitigate or manage that risk. Further discussion about mitigations strategies can be found in Chapter 11. The risk assessor should implement these mitigation strategies and integrate them into the project's overall risk management plan.

The sixth step is to monitor the implemented mitigation strategies. The risk assessor should frequently monitor the AI system and its interaction with synthetic biology applications frequently and look for any emerging risks or changes in the risk profile.

TABLE 4.4
Description of the Seven Degrees of Automation for Artificial Intelligence Systems

Level of Automation	System	Degree of Human Control	System Control	Risk Level
1. No automation	Heteronomous	Full	The operator is in full control	Low
2. Assistance	Heteronomous	Full	The system assists an operator	Low
3. Partial automation	Heteronomous	Full	The system has some functions that are fully automated, but it remains under the control of an external agent/operator	Moderate
4. Conditional automation	Heteronomous	Full	The system's automation is specific and sustained, but an external operator is supervising and ready to take control when necessary	Moderate
5. High automation	Heteronomous	Full	The system performs parts of its mission without external intervention	Moderate
6. Full automation	Heteronomous	Partial	The system performs its entire mission without external intervention	High
7. Autonomy	Autonomous	None	The system performs and modifies its operation domain or its goals without external intervention, control, or oversight	Very high

These seven degrees of automation[112] for Artificial Intelligence systems include: (1) no automation, (2) assistance, (3) partial automation, (4) conditional automation, (5) high automation, (6) full automation, and (7) autonomy.[110] For each level, this table assesses the degree of human control, system control, and the associated risk level for misuse.

The seventh step is to regularly review and update the risk assessment based on changes in the AI system, the regulatory landscape, or any new relevant information.

Example: The author has previously published a specific detailed example of how to apply these tables to a large language model (LLM) application in synthetic biology biosecurity risk assessment.[90]

4.9 OVERSIGHT IMPLICATIONS AT A GLOBAL AND NATIONAL SCALE

AI and synthetic biology are indeed driving a revolution in biomedicine, opening new possibilities for treating diseases, improving health, and understanding fundamental aspects of biology. However, their potential also brings significant biosecurity implications. To ensure that these technologies are used safely and effectively, proactive and comprehensive oversight at both national and global scales is crucial:

TABLE 4.5
Risk Assessment Guidelines for Conducting an AI Biorisk Assessment in Synthetic Biology

AI Model Type	Risk Identification (Risk Introduced by Using AI Tools)	Consequences	Risk Level (Compared to Using a Search Engine)	Mitigation	Residual Risk (After Mitigation is Implemented)
Large language models (LLMs)	Increased access to knowledge and capabilities (over a search engine)	Lowers the barrier to biological misuse	Low		Low
	Teaching about dual-use topics can answer specific and relevant questions about biological weapons and development	Enable smaller/amateur BW efforts to overcome key technical bottlenecks without overusing resources or without drawing too much attention to themselves	Moderate	1. Pre-release model evaluation of the biosecurity risks: • External and independent audit of the models • Conduct a structured set of tests to evaluate the vulnerabilities in the model. This is not a trivial task to accomplish because the system being a "black box". It is currently not a feasible theoretical framework, but an aspirational one • Test the model's ability to plan a possible biological attack • Use tooling and fine-tuning of the models to evaluate system modifications and new capabilities after system modifications	Low
	Ease of identification of biological misuse	Help non-experts identify specific pathogens and targets for misuse and guide them on how to design agents tailored to a specific goal	Moderate		Low
	Instruction and troubleshooting	As AI lab assistants become more effective, they can help troubleshoot experiments and give personalized instructions to be used for malicious intent	High		Low

(*Continued*)

TABLE 4.5 (*Continued*)
Risk Assessment Guidelines for Conducting an AI Biorisk Assessment in Synthetic Biology

AI Model Type	Risk Identification (Risk Introduced by Using AI Tools)	Consequences	Risk Level (Compared to Using a Search Engine)	Mitigation	Residual Risk (After Mitigation is Implemented)
	Autonomous science	As capabilities to generate scientific work with minimal or non-existent human input, it will become easier to coordinate automated BW programs under secrecy	Very high	2. Do a risk-benefit analysis of unrestricted access and security: • Collect data to determine the adequate level of access needed to prevent misuse without stifling progress	High
Biological design tools (BDTs)	Dual-use concerns	Design and create proteins or whole organisms optimized across different functions for misuse	Moderate	3. Implement mandatory gene synthesis screening: • Improve screening tools to keep pace with biodesign	Low
	Increased worst-case scenarios	Increased ceiling of capabilities by designing new pathogens that bypass the trade-off between transmissibility and virulence	Moderate	• Implement protein structure predictor capabilities • Compare not only gene sequences and taxonomy but also structural predictions to other harmful toxins/proteins/genes	Low
	Making agents more predictable and targetable	Increase the harm ceiling by designing new agents more specific to certain geographical areas or populations	High	• There is a very low barrier to entry here and it is already being implemented by companies. It is important to verify that this is being done	Low

(*Continued*)

TABLE 4.5 (*Continued*)
Risk Assessment Guidelines for Conducting an AI Biorisk Assessment in Synthetic Biology

AI Model Type	Risk Identification (Risk Introduced by Using AI Tools)	Consequences	Risk Level (Compared to Using a Search Engine)	Mitigation	Residual Risk (After Mitigation is Implemented)
	Sequence design	Easier design of harmful agents, bypassing taxonomy, and sequence similarity-based controls	High		Moderate
Both models combined (LLMs and BDTs)	The combined tool will augment the risks of both independent AI model types	Elevated risk of all the consequences mentioned above in this table	Very high		High

The table outlines a process for identifying risks associated with different AI model types, their potential consequences, the corresponding risk levels, and recommended mitigation strategies.

- **Being Proactive in Anticipating Gaps and Addressing Them Early**: As AI and synthetic biology evolve, it is vital to anticipate potential biosecurity gaps and address them early. This could involve conducting foresight exercises to anticipate future developments and their implications, building security measures from the outset (the principle of "security by design"), and establishing mechanisms for ongoing monitoring and risk assessment. Early engagement with diverse stakeholders, including researchers, policymakers, biosecurity experts, and the public, can help ensure that these measures are comprehensive and widely accepted.
- **Importance of the Technology**: The potential of AI and synthetic biology to drive advancements in biomedicine is immense. From developing personalized treatments and novel vaccines to engineering healthier and more sustainable food, these technologies could bring significant health, economic, and societal benefits. Given their importance, it is crucial that we find ways to harness these technologies while managing their risks. This requires not just strong oversight, but also sustained investment in research, training, and infrastructure.
- **National and Global Scales**: Oversight of AI and synthetic biology needs to occur at both national and global scales. At the national level, countries need to establish regulations and guidelines for using these technologies,

invest in the necessary infrastructure and expertise, and promote a culture of responsibility and ethics in their use. However, given the global nature of these technologies and their potential impacts, international cooperation is also essential. This could involve sharing best practices, harmonizing regulations, collaborating on risk assessments, and jointly responding to biosecurity incidents. International organizations and think tanks can play a key role in facilitating such cooperation.

To sum up, effective oversight of AI and synthetic biology requires proactive and ongoing efforts at both national and global scales. By anticipating and addressing biosecurity gaps early, we can ensure that these technologies are used safely and effectively to drive the next revolution in biomedicine.

4.10 LEGAL AND ETHICAL CONSIDERATIONS OF AI USAGE

As with any rapidly advancing technology, the integration of AI in synthetic biology raises numerous legal and ethical concerns, including issues of access, privacy, accountability, and impacts on society. Some of these concerns include:

- **Fair and Equitable Access to AI Technology Globally**: AI has the potential to significantly advance synthetic biology, including in areas like disease diagnosis, drug discovery, and environmental conservation. However, the benefits of these advancements will be unevenly distributed if access to AI technology is not equitable. High costs, lack of infrastructure, and lack of trained personnel can limit access to AI in many parts of the world, particularly in low- and middle-income countries. This could exacerbate existing health and socioeconomic disparities. Efforts are needed to democratize access to AI, such as through capacity-building initiatives, open-source software, and collaborations between high-income and low-income countries.
- **Ethical Use of the Technology**: The intersection of AI and synthetic biology brings along a complex web of ethical considerations. These technologies have the potential to alter life as we know it, opening possibilities that could fundamentally transform medicine, agriculture, and environmental science. However, alongside these potential benefits are significant ethical concerns that must be addressed:
  - One of the primary ethical considerations is the appropriate use of technology. As AI and synthetic biology have advanced, we now have the capacity to alter genomes, create synthetic organisms, and design biological systems in ways that were not previously possible. The question is not only what we can do, but what we should do. There are limits that society may wish to place on these technologies due to concerns about unforeseen consequences of fundamentally altering what it means to be human.
  - For instance, gene-editing technologies like CRISPR can potentially be used to edit the human germline, making changes that could be passed down to future generations. This raises ethical questions about consent (can future generations consent to the changes we make now?), equity

(who gets to decide and benefit from such technology?), and uncertainty (what are the long-term effects?).
    - Using AI in synthetic biology also raises concerns about consent. AI systems often rely on large datasets for training and inference. In synthetic biology, these could be genomic or health-related data, which are often sensitive and personal. Consent procedures for data use, particularly big data, and AI need to be robust and clear to individuals. People should have the right to understand how their data will be used and to make an informed choice about whether to participate.
    - The development and use of these technologies also have implications for social justice and equity. If only a privileged few can access the benefits of AI and synthetic biology (such as personalized medicine or enhanced crops), it could exacerbate existing inequalities. Ethical use of the technology must therefore include considerations of fair access and distribution.
    - Finally, there is the issue of transparency and public engagement. For such powerful technologies, there should be a public dialogue about their development and use. This includes not only clear communication of the potential benefits and risks, but also opportunities for public input into decision-making processes. The ethical use of AI and synthetic biology should be a matter of societal decision-making, rather than being driven solely by scientists or industry.
- **Accountability**: As the interplay of AI and synthetic biology continues to evolve and expand, accountability becomes more complex. There are multiple stakeholders involved, each playing distinct roles in the application of these technologies. This includes AI developers, synthetic biologists, researchers using the technologies, institutions overseeing the work, and regulators. Establishing clear lines of accountability is necessary to ensure that these technologies are used responsibly and that any harmful outcomes can be properly addressed:
    - Consider the scenario where a genetically engineered insect, created to combat a disease like malaria, unexpectedly causes ecological disruption – such as the decimation of a local lizard population that depended on the mosquitos for food. Pinpointing responsibility and liability in this case is multifaceted.
    - The developers of the AI system used in designing the gene-editing solution could be one point of responsibility. Did they incorporate all necessary safeguards and bias mitigation strategies in their algorithms? Was their system designed to consider all potential ecological effects, and if not, should it have been?
    - Researchers who employed the technology might also be held accountable. Did they perform comprehensive threats and vulnerabilities assessment before releasing the genetically engineered insect? Did they consider all potential off-target effects and ecological disruptions, and follow all guidelines and regulations in their work?
    - The institutions or companies that supported the work could also be held liable. Did they provide adequate oversight and enforce necessary

safety and ethical guidelines? Did they have contingency plans for unforeseen ecological impacts?
  - Finally, regulators and policymakers play a significant role. They are tasked with establishing the legal and regulatory frameworks that guide such work. If those frameworks are inadequate or fail to foresee potential risks, there may be shared accountability as well.
  - Accountability may be distributed across several entities. The key is having clear guidelines, rigorous oversight mechanisms, and robust legal frameworks in place before the implementation of such projects. This way, if something goes wrong, there is a clear understanding of the roles, responsibilities, and potential legal repercussions.
  - The law often lags behind technology, and current legal systems may not be fully equipped to handle these scenarios. Therefore, there is a pressing need for legal and regulatory innovation to keep pace with advances in AI and synthetic biology. This should involve continuous dialogue among scientists, ethicists, lawyers, policymakers, and the broader public. Moreover, international cooperation will be crucial given the global implications of these technologies.
- **Impacts on Society**: The ethical and legal ramifications of such a scenario involving synthetic biology and biosecurity are vast and complex. As we increasingly use synthetic biology to engineer solutions like biofuel-producing algae, the potential for unintended negative impacts on society and the environment rises. These impacts can be both local and global, affecting communities, ecosystems, and economies:
  - For example, if a European Union-based multinational company were to place large algae ponds in a low-income country and a subsequent spill caused significant environmental and societal harm, determining liability and responsibility would involve several considerations.
  - First, the company would have primary responsibility, particularly if it failed to implement appropriate safety measures and risk mitigation strategies. Companies conducting these types of activities are expected to follow best practices for safety and to carry out thorough risk assessments. If these practices are ignored or poorly implemented, the company can be liable for the consequences.
  - However, responsibility also extends beyond the company to the regulatory bodies and policymakers of both the home country (EU in this case) and the host country. The host country's government has the duty to safeguard its environment and citizens by ensuring that foreign and domestic companies adhere to safe practices and comply with environmental regulations. If there were lapses in the regulatory oversight that contributed to the incident, the government might share in the liability.
  - Similarly, the EU, as the home base of the multinational company, has an ethical responsibility to ensure that its companies behave responsibly overseas. This means enforcing stringent corporate standards and pursuing legal action if a company causes harm abroad.

- Given the cross-border nature of such incidents, international laws, and regulations, such as those set by the United Nations or other international bodies, may also come into play. These bodies could potentially mediate the dispute, assess liability, and implement sanctions if needed.
- In addition to these legal and regulatory considerations, ethical responsibility could be shared among the scientific community involved in developing the technology. The developers of synthetic biology techniques should work to understand potential risks and to build safety features into their systems, as well as to communicate these risks clearly to users and regulators.
- This scenario highlights the need for robust international guidelines and cooperation on biosecurity. It also emphasizes the necessity for corporations to not only adhere to regulatory guidelines, but also commit to ethical best practices, including thorough risk assessment, implementation of effective safety measures, transparency in their operations, and engagement with local communities. The local communities affected should also be part of the conversation, as they have a right to engage in decision-making processes that impact their lives and the environment.

Addressing the legal and ethical concerns of integrating AI in synthetic biology requires a multifaceted approach, involving not just the development of new laws and regulations, but also fostering a strong culture of ethics and responsibility in the use of these technologies, as well as ongoing dialogue with all stakeholders, including the public.

4.11 CONCLUDING THOUGHTS

AI is transforming synthetic biology and biosecurity, introducing both unprecedented innovation and significant risks. AI enhances biomedical research and therapy development through advanced data analysis, accelerating problem-solving and design optimization. However, it also raises biosecurity issues, including the misuse for creating biological threats and data breaches. Addressing these concerns necessitates strong biosecurity protocols, ethical standards, and global cooperation to ensure AI's ethical application. Moreover, AI's role in advancing public health and biomedical research is undeniable, yet it brings ethical and biosecurity challenges, highlighting the need for international dialogue and policymaking. The chapter provides a detailed biosecurity risk assessment framework for AI in synthetic biology, advocating for a balanced approach to technological advancement and safety. It underscores the importance of a collaborative, multidisciplinary effort to navigate the benefits and risks of AI, involving a wide range of stakeholders in the process.

4.12 KEY TAKEAWAYS

- **AI in Synthetic Biology**: Enhancing research capabilities and accelerating discoveries in gene editing, protein design, and drug development.
- **Biosecurity Concerns**: Addressing the potential misuse of AI for creating harmful biological agents and the need for robust security measures.

- **Ethical Implications**: Highlighting ethical dilemmas in AI applications, such as data privacy and equitable access to technology.
- **Global Collaboration**: Emphasizing the importance of international cooperation in addressing biosecurity risks associated with AI.
- **Data Management and Analysis**: Leveraging AI for efficient handling and interpretation of large-scale biological data.
- **Public Health Surveillance**: Utilizing AI in monitoring and responding to health threats, including disease outbreak prediction.
- **AI Risk Assessment Framework**: Presenting a set of innovative tools and methodology for comprehensive risk assessment using a structured approach to identifying specific AI technologies, assessing risks and vulnerabilities, and implementing effective mitigation strategies.

4.13 THOUGHT-PROVOKING QUESTIONS

4.1 What policies can be implemented to ensure AI's proper transparency in decision-making processes and prevent its potential for misuse?

4.2 How can we develop AI tools in synthetic biology that not only recognize and prevent threats, but also guarantee that such innovations contribute positively to society? What comprehensive biosecurity protocols are essential to avert the abuse of AI's dual-use potential?

4.3 What strategies are necessary to harmonize the advantages of AI innovations with the requirements for ethical supervision and regulatory control? How can these strategies accommodate ethical considerations pertaining to the openness and responsibility of AI systems, the possibility of misuse, and the wider consequences of synthesizing life forms?

4.4 What parameters/elements should be included in developing a model for AI use in synthetic biology? The model must also be adaptable to keep pace with advancements in these rapidly evolving fields.

4.5 How can we ensure that the integration of AI in synthetic biology is done in a manner that safeguards responsible use?

4.6 What regulatory frameworks might be necessary to oversee the use of AI in synthetic biology?

4.7 How can ethical guidelines be developed and implemented to govern the intersection of AI and synthetic biology?

4.8 How might AI advancements shape synthetic biology's future, and what biosecurity implications could this entail?

4.9 What are some potential challenges in integrating AI into synthetic biology, and how might they be addressed to ensure an adequate level of biosecurity?

4.10 Who ought to bear the responsibility for establishing, implementing, and enforcing AI regulations at a regional level?

4.11 In the scenario where a European Union-affiliated corporation installs extensive algae ponds in a developing nation, resulting in an environmentally and socially detrimental spill, various factors must be considered to determine accountability. Which entity should have provided oversight and authorization for such a project – the company's country of origin, the nation hosting the operation, and should it be managed at a local or regional level?

4.12 As AI is integrated into synthetic biology, what are the gaps in the current regulatory frameworks and insufficient understanding of AI's potential misuse that could prevent responsible development?

4.13 What other strategies should be included in developing a robust AI biosecurity protocol while fostering global cooperation and enhancing public awareness of biosecurity risks?

4.14 What collaborative risk assessment exercises and shared development of mitigation strategies can enhance stakeholder involvement about AI-related risks in synthetic biology? How can the biosecurity community better engage with stakeholders to address this issue?

4.15 What other key factors should be considered in determining the biosecurity risk of AI applications in synthetic biology?

4.16 What level of expertise and ease of access to AI tools would be required for the misuse (such as dual use or delivering a successful attack) of AI in synthetic biology?

4.17 What new measures should be included in updated methodologies to address biosecurity concerns about new complex and dynamic threats arising from AI's capabilities and vulnerabilities?

5 Global Regulatory Frameworks and Policies for Synthetic Biology

Around the world, there are various unharmonized regulatory frameworks and policies governing synthetic biology. This field's state of regulation varies significantly from region to region. Some areas boast well-established and comprehensive regulatory structures, mainly focusing on biosecurity aspects. In contrast, other regions are still nascent in developing and implementing their regulatory guidelines. One key challenge highlighted in this chapter is harmonizing these regulations across international borders. Given the inherently global nature of scientific research and the collaborative essence of synthetic biology, this lack of standardization poses significant hurdles. The chapter emphasizes the need for regulatory frameworks that are robust and comprehensive but also flexible and adaptive. Such frameworks must be capable of evolving alongside the rapid advancements in synthetic biology, ensuring safety and ethical integrity while also fostering societal acceptance and support.

DOI: 10.1201/9781003423171-5

This chapter underscores the critical importance of developing dynamic regulatory frameworks that can effectively respond to the fast-paced changes in synthetic biology. It calls for a concerted effort among international stakeholders to bridge regulatory gaps, facilitating responsible innovation and collaboration in this burgeoning field.

5.1 INTERNATIONAL PERSPECTIVES ON BIOSECURITY

As the capabilities of synthetic biology advance, so too does the need for effective biosecurity measures to prevent the misuse of these technologies. The perspectives on biosecurity can differ significantly across countries and cultures, reflecting differing priorities, regulatory philosophies, and cultural norms. However, some key points of consideration are relevant on both national and international scales, including:

- **Regulation and Oversight**: Ensuring appropriate oversight of synthetic biology research and application is critical to mitigate risks. At the national level, this may include creating or harmonizing legislation, standards, and regulatory bodies. Internationally, it might involve treaties, conventions, or other agreements to establish shared norms and expectations, such as the Biological Weapons Convention.
- **Transparency and Information Sharing**: Transparency in research, development, and application of synthetic biology is crucial for maintaining trust among nations and stakeholders. This can be challenging as it must be balanced with the need to protect proprietary information and prevent misuse of potentially dangerous knowledge.
- **Collaboration and Capacity Building**: Countries can have varying capacities to engage in and regulate synthetic biology. International collaboration can help to build capacity (to perform their own research) in countries where it is lacking, ensure more equitable access to the benefits of synthetic biology, and foster a shared approach to managing its risks.
- **Dual-Use Research Concern**: Some areas of research in synthetic biology can be considered "dual use" – they have the potential for both beneficial applications and misuse for harmful purposes. Identifying and managing dual-use research is a significant challenge that requires careful risk-benefit analyses and may involve national and international approaches.
- **Education and Training**: To ensure biosecurity, it is important to educate those working in synthetic biology about potential biosecurity risks and their ethical and professional responsibilities. This can be done through degree programs, professional training, and continuing education.
- **Engaging the Public**: Public engagement can help to ensure that societal values and concerns are incorporated into decisions about synthetic biology and biosecurity. This can be done through public consultations, citizen juries, public dialogues, and other participatory mechanisms.

Different countries may emphasize different elements of these strategies depending on their specific circumstances and values. However, due to the global nature of both the potential benefits and risks of synthetic biology, international cooperation will be key to ensuring effective mitigation measures in biosecurity.

5.1.1 Biological and Toxin Weapons Convention (BTWC)

The Biological and Toxin Weapons Convention (BTWC) is an international treaty that categorically bans the development, production, acquisition, transfer, stockpiling, and use of biological and toxin weapons. It was opened for signature in 1972 and has since played a significant role in establishing a solid norm against biological warfare.[114] As of publication of this book, 183 countries have signed this treaty, 4 countries are signatory, and 10 countries have neither signed nor ratified the document.[115] Key points relevant to synthetic biology in BTWC are:

- **Scope of Prohibition**: BTWC's prohibition covers not only biological agents but also the weapons that might deliver such agents. Article I of the BTWC prohibits the development, production, stockpiling, or acquisition of biological agents or toxins of types and quantities without justification for prophylactic, protective, or other peaceful purposes. Synthetic biology, which could potentially be used to create or modify such agents, falls within the scope of the Convention.
- **Oversight of Research**: The BTWC obliges states to prohibit and prevent the misuse of biological science and technology in their jurisdiction. This has implications for synthetic biology research and application. It means that states must ensure that the research and use of synthetic biology for peaceful purposes do not inadvertently contribute to the production or enhancement of biological weapons.
- **Preventing the Misuse of Science**: The BTWC provides a framework for member states to discuss and collaborate on preventing the misuse of science to produce biological weapons. The advent of synthetic biology, which enables the construction of entirely new biological systems, underscores the importance of such collaborative efforts to prevent misuse.
- **Confidence-Building Measures**: Under the BTWC, member states are encouraged to submit confidence-building measures, including information on high-risk research and high-containment laboratories within their jurisdiction. Synthetic biology, with its ability to construct or modify biological agents, is a field where such confidence-building measures would be particularly relevant.
- **Dual-Use Nature of Research**: Like many areas of scientific inquiry, synthetic biology is "dual use" – its methods and findings can be used for beneficial and harmful purposes. This raises significant ethical and safety issues, and it is where BTWC's oversight and regulatory measures are highly relevant.

The principles and framework of BTWC are very pertinent to synthetic biology. As the field continues to advance, the international community needs to apply and adapt these regulations to ensure that synthetic biology is used responsibly and does not pose biosecurity threats.

5.1.2 Cartagena Protocol on Biosafety

The Cartagena Protocol on Biosafety is an international agreement aimed at ensuring the safe handling, transport, and use of living-modified organisms (LMOs) that are created through modern biotechnology.[116] The protocol protects biological diversity and human health from the potential risks associated with LMOs. It was adopted in 2000 as a supplementary agreement to the United Nations Convention on Biological Diversity (CBD). Several aspects of the Cartagena Protocol on Biosafety that are relevant to synthetic biology include:

- **Advanced Informed Agreement (AIA) procedure**: The protocol established an AIA procedure for ensuring that countries are provided with the information necessary to make informed decisions before agreeing to importing LMOs intended for intentional introduction into the environment.
- **Risk Assessment and Risk Management**: The protocol sets out principles and methodologies for the assessment and management of potential biosafety risks. These include conducting a case-by-case scientific risk assessment before any LMO is approved for use and establishing appropriate risk management strategies to ensure that the use of LMOs does not lead to adverse effects on biodiversity or human health.
- **Identification and Documentation**: The protocol requires that shipments of LMOs are clearly identified as such, and accompanied by appropriate documentation specifying, among other things, the identity of the LMOs, the contact point for further information, and any requirements for their safe handling, storage, transport, and use.
- **Biosafety Clearinghouse (BCH)**: The protocol established a BCH as a mechanism for facilitating the exchange of scientific, technical, environmental, and legal information on LMOs, as well as for assisting with implementing the Protocol's provisions.
- **Socioeconomic Considerations**: Parties to the Protocol may take socioeconomic considerations in their decision-making about LMOs, particularly regarding the value of biodiversity to Indigenous and local communities.
- **Public Participation and Confidentiality**: The Protocol highlights the importance of public participation in decision-making processes related to LMOs and stipulates that biosafety information should be made public while protecting confidential business information.

In synthetic biology, the Protocol serves as a crucial international regulatory framework to guide the safe development and application of novel organisms. However, its effectiveness depends on how well its provisions are implemented and enforced by each party country.

5.1.3 The Nagoya Protocol on Access and Benefit Sharing

The Nagoya Protocol on Access and Benefit Sharing, also known as simply the "Nagoya Protocol," is an international agreement that aims to fairly and equitably share the benefits arising from the utilization of genetic resources.[117] Adopted in

2010, it supplements the United Nations Convention on Biological Diversity (CBD) and carries significant implications for synthetic biology. Some key points of the Nagoya Protocol, as they pertain to synthetic biology, include:

- **Access to Genetic Resources**: The Nagoya Protocol requires that access to genetic resources be based on prior informed consent and mutually agreed terms between the provider and the user. In synthetic biology, this has significant implications for the sourcing of genetic material that may be used in engineering new biological systems or entities.
- **Benefit-Sharing**: The Protocol stipulates that those benefits arising from the use of genetic resources, such as commercial profits from new drugs or therapies, should be shared fairly and equitably with the country providing the resources. In synthetic biology, this could apply to genetically engineered products derived from these resources.
- **Traditional Knowledge**: The Protocol recognizes the importance of traditional knowledge associated with genetic resources and establishes provisions for sharing benefits with Indigenous and local communities when their knowledge is used.
- **Compliance Measures**: The Protocol introduces obligations for its Parties to develop mechanisms to promote compliance with its provisions. This includes ensuring that genetic resources utilized within their jurisdiction have been accessed in accordance with prior informed consent and that mutually agreed terms have been established.
- **Global Multilateral Benefit-Sharing Mechanism**: The Nagoya Protocol introduces the potential for a global multilateral benefit-sharing mechanism to address situations where it is not possible to grant or obtain prior informed consent. This may include cases where genetic resources and traditional knowledge are found in transboundary situations or held by Indigenous and local communities.

In synthetic biology, these provisions can be quite significant. For instance, the creation of synthetic organisms could involve using genetic resources obtained from different parts of the world, raising questions about access and benefit-sharing under the Nagoya Protocol. Implementation and enforcement of the protocol can ensure that synthetic biology is conducted in a manner that respects the rights of countries and communities providing genetic resources and that benefits are shared fairly and equitably.

5.1.4 World Health Organization (WHO) Laboratory Biosafety Manual

The WHO Laboratory Biosafety Manual Fourth Edition (2022), supported by seven monographs, provides global guidelines for the safe handling and containment of biological agents, including those created or modified through synthetic biology, in laboratory environments.[118] As a reference document adopted by numerous countries to create their own national biosafety guidelines, the manual delineates best practices to minimize the risk of accidental release or exposure to harmful biological agents.

Its main objective is to guide personnel in the handling and containing a range of microorganisms, particularly those that are pathogenic or potentially pathogenic, thereby ensuring safety and security in laboratories worldwide. Potential points relevant to synthetic biology include:

- **Biosafety Risk Assessment**: Synthetic biology often involves the manipulation of biological agents, including those that might be pathogenic. The principles of biosafety risk assessment in the manual would therefore be highly relevant.
- **Containment Levels**: In the fourth edition, the manual changes approach and moves away from outlining biosafety levels (BSL1–BSL4) of containment and toward containment based on the biohazard risk (core control measures, heightened control measures, and maximum containment measures), which guide the handling of organisms engineered through synthetic biology.
- **Use of Personal Protective Equipment (PPE)**: The manual's guidance on using PPE would be relevant when working with engineered organisms.
- **Waste Disposal**: Proper disposal methods for biological waste, as outlined in the manual, would apply to waste generated in synthetic biology labs.
- **Emergency Response**: Guidelines for safely transporting biological materials might also apply to materials engineered using synthetic biology.
- **Transportation**: Guidelines for the safe transport of biological materials might also apply to materials engineered using synthetic biology.
- **Training**: The manual emphasizes the importance of training laboratory staff in biosafety, which is crucial in a synthetic biology context as well.

5.1.5 Consideration on the Status of International Efforts in Biosecurity

The international perspectives on biosecurity are as diverse as the countries they originate from, reflecting various cultural norms, regulatory approaches, and prioritized objectives. However, the underlying global consensus insists upon robust regulatory oversight, transparency, and information sharing. It also calls for international collaboration and capacity building to ensure adequate mitigation measures are implemented, particularly with the emerging capabilities of synthetic biology. Countries may place differing emphases on these strategies according to their unique circumstances, but international cooperation is crucial due to the potential global impacts of synthetic biology.

The BTWC, a major international treaty, has played a significant role in establishing norms against biological warfare. Synthetic biology, with the potential to modify biological agents, falls within the scope of the BTWC, prompting states to ensure that peaceful synthetic biology research does not inadvertently contribute to the creation of biological weapons. Similarly, the Cartagena Protocol on Biosafety ensures the safe handling, transport, and use of LMOs, guiding the safe development and application of novel organisms produced by synthetic biology.

Equally important is the Nagoya Protocol on Access and Benefit Sharing, which promotes the fair and equitable distribution of benefits arising from the utilization of genetic resources. This protocol has notable implications for synthetic biology, which often involves using genetic resources from various parts of the world.

Lastly, the WHO Laboratory Biosafety Manual provides global guidelines for safely handling biological agents in laboratory environments, offering key insights relevant to synthetic biology.

While individual countries may prioritize different elements of these strategies according to their circumstances, these international agreements serve as a collective guide toward ensuring effective biosecurity in synthetic biology. With the potential global impacts of synthetic biology, cooperation and alignment with these international standards are important.

5.2 UNITED STATES' PERSPECTIVES ON BIOSECURITY

5.2.1 Legal Framework for Biosecurity: The United States' Prohibition of Biological Weapons

Within the United States, biosecurity is governed by stringent laws that underscore the country's commitment to mitigating the risks associated with biological agents and toxins. The 18 U.S. Code § 175[119] explicitly prohibits the development, production, stockpiling, transfer, acquisition, retention, or possession of biological agents, toxins, or delivery systems intended for use as weapons. This law reflects a comprehensive approach to biosecurity, encompassing the outright use of such materials as weapons and their development and possession with the intent of weaponization.

The law is categorical in penalizing any individual or entity that knowingly engages in activities related to biological weapons, with the possibility of severe penalties, including life imprisonment. Significantly, it extends extraterritorial jurisdiction, implying that U.S. nationals can be held accountable for such offenses regardless of where they occur. Additionally, the law acknowledges the legitimate use of biological agents and toxins for prophylactic, protective, bona fide research, or other peaceful purposes, drawing a clear distinction between lawful scientific pursuits and activities with malicious intent.

This legislation is a cornerstone in the United States' efforts to prevent the misuse of biological materials that could threaten national and international security. It underscores the delicate balance that must be maintained between advancing scientific research and safeguarding against the potential misuse of scientific discoveries. The law aligns with international efforts and norms aimed at preventing the proliferation of biological weapons, reflecting a global consensus on the need for rigorous controls over hazardous biological materials. This law is a legal framework and statement of ethical responsibility, guiding the scientific community in responsible conduct and contributing to global biosecurity efforts.

5.2.2 U.S. Department of Defense Biodefense Posture Review (BPR)

In 2023, the U.S. Department of Defense issued a biodefense posture review (BPR),[120] which outlines significant reforms that deter the use of bioweapons, aid in the ability to respond to natural outbreaks rapidly, and minimize the global risk of laboratory accidents. This review was borne out of COVID-19 response efforts and takes a long-view approach considering recent developments in biotechnology, specifically synthetic biology, which are driving an increase in the scope and diversity of biothreats. They also consider the emerging and re-emerging infectious diseases of high consequence that are expected to develop and spread more frequently due to the planet's climate change, population growth, and global movements. The most important activities outlined in this review are as follows:

- Expand capabilities to understanding and be aware of biothreat.
- Innovate and modernize biodefense capabilities for the next 12 years to maintain readiness and resiliency.
- Improve readiness through training and exercising to aid the modernization efforts.
- Establish the Biodefense Council to synchronize, coordinate, and integrate authorities and responsibilities to provide an empowered and collaborative approach to sustained biodefense.

This review assessed the biodefense landscape in the U.S. Armed Forces and provided actionable and practical recommendations to bolster the country's biodefense capability. It offers a clear pathway to close gaps and vulnerabilities identified during the COVID-19 pandemic response. However, it is only focused on the armed forces. This report is so important that it should be expanded and adapted to non-military-related activities (academic, entrepreneurial, and clinical) that could benefit from more explicit guidance and support set forth by the government's authority, especially in light of a recent incident involving a secret biomedical laboratory in California that is still under investigation.[121]

5.2.3 Biosafety in Microbiological and Biomedical Laboratories (BMBL)

The Biosafety in Microbiological and Biomedical Laboratories (BMBL) is a comprehensive guide to biosafety in microbiology and biomedical laboratories.[122] The BMBL was first published in 1984 and has been updated several times. The sixth edition of the BMBL was published in 2020. The BMBL guides on a wide range of biosafety topics, including:

- The classification of biological agents according to their potential risk (risk group).
- The four containment levels required for different classes of biological agents.
- The training and responsibilities of laboratory personnel.
- The procedures for handling, storage, and disposal of biological agents.
- The design and operation of biosafety laboratories.
- The response to biosafety incidents.
- A special section for large-scale production of biologicals.

The main conclusion of the BMBL is that the safe and responsible conduct of research involving biological agents requires a comprehensive biosafety program. The BMBL provides a framework for developing and implementing such a program. Some additional key points from the BMBL include:

- It is an advisory document, not a regulatory document. However, many countries and organizations have adopted the BMBL as a basis for their biosafety regulations.
- It is intended to be used with other biosafety guidance documents, such as the NIH Guidelines for Research Involving Recombinant or Synthetic Nucleic Acid Molecules.
- It is subject to periodic review and update. The most recent update was published in 2020.

The BMBL is an essential resource for anyone involved in the safe and responsible conduct of research involving biological agents. The BMBL provides a comprehensive framework for developing and implementing a biosafety program, and it is an invaluable tool for protecting researchers, laboratory workers, the public, and the environment. In addition to the main conclusion, some other key takeaways from the BMBL include:

- Biosafety is an essential component of any laboratory that works with biological agents.
- The level of biosafety mitigation measures required depends on the risk posed by the biological agents being worked with.
- Biosafety programs should be tailored to the specific needs of the laboratory and the specific work to be performed.
- Biosafety training should be provided to all laboratory personnel.
- Biological agents should be handled and stored safely.
- Biosafety incidents should be reported and investigated promptly, and mitigation measures implemented to prevent them from happening again in the future.

The BMBL is a valuable resource for anyone in a laboratory that handles biological agents. The BMBL guides a wide range of biosafety topics, and it is an essential tool for protecting researchers, laboratory workers, the public, and the environment.

5.2.4 The Ethics of Synthetic Biology and Emerging Technologies (2010)

The Presidential Commission for the Study of Bioethical Issues (PCSBI) report "New Directions: The Ethics of Synthetic Biology and Emerging Technologies" was published in 2010 and examines the ethical implications of synthetic biology, a new field of biotechnology that combines engineering principles with existing biotechnology techniques, such as DNA sequencing and genome editing, to modify organisms or create new ones.[123] The report identifies several ethical issues raised by synthetic biology, including:

- The potential for synthetic biology to create new pathogens or toxins.
- The potential for synthetic biology to create new forms of discrimination or social exclusion.
- The potential for synthetic biology to disrupt the natural world.
- The potential for synthetic biology to be used for military purposes.

The report also discusses a number of ethical principles that can guide the development of synthetic biology, including:

- The principle of beneficence requires that scientists act in the best interests of humanity.
- The principle of nonmaleficence, which requires that scientists avoid causing harm.
- The principle of justice requires that the benefits and risks of synthetic biology be distributed fairly.
- The principle of respect for autonomy requires that scientists respect the rights and choices of individuals.

The main conclusion of the report is that synthetic biology is a powerful new technology that has the potential to benefit humanity in many ways. However, the report also warns that synthetic biology could be used for harmful purposes, and it calls for developing of ethical guidelines to ensure that the technology is used responsibly. Some additional key points from the report include:

- The report defines synthetic biology as "the design and construction of new biological entities, components, or systems, or the redesign of existing biological systems, to achieve specific goals."
- The report identifies three main categories of ethical issues raised by synthetic biology: (1) the potential for misuse, (2) the potential for unintended consequences, and (3) the need for responsible governance.
- The report discusses a number of ethical principles that can guide the development of synthetic biology, including beneficence, nonmaleficence, justice, and respect for autonomy.
- The report calls for developing ethical guidelines for synthetic biology, and it recommends that these guidelines be developed inclusively and transparently.

The PCSBI report is an important contribution to the debate on the ethical implications of synthetic biology. The report provides a comprehensive overview of the ethical issues raised by this technology, and it offers a set of ethical principles that can guide its development. It is a valuable resource for scientists, policymakers, and the public as they seek to ensure that synthetic biology is used responsibly and for the benefit of humanity.

5.2.5 NIH Guidelines for Research Involving Recombinant or Synthetic Nucleic Acid Molecules

The NIH Guidelines for Research Involving Recombinant or Synthetic Nucleic Acid Molecules, or simply the "NIH Guidelines," are a set of biosafety and biosecurity regulations governing research involving recombinant or synthetic nucleic acid molecules in the United States.[124] The guidelines were initially issued in 1976 and have been updated several times since then. In April, 2024 the guidelines were updated to include guidance on gene drive experiments.

The NIH Guidelines are designed to protect researchers, laboratory workers, the public, and the environment from the potential risks of recombinant or synthetic nucleic acid molecules. The guidelines cover a wide range of topics, including:

- The classification of recombinant or synthetic nucleic acid molecules according to their potential risk.
- The containment requirements for different classes of recombinant or synthetic nucleic acid molecules.
- The training and responsibilities of laboratory personnel.
- The procedures for handling, storing, and disposing of recombinant or synthetic nucleic acid molecules.
- The review and approval of recombinant or synthetic nucleic acid molecule experiments.

The main conclusion of the NIH Guidelines is that recombinant or synthetic nucleic acid molecules can pose a variety of risks, but that these risks can be managed by implementing appropriate biosafety and biosecurity mitigation measures. The guidelines provide a comprehensive framework for managing these risks, and they are an essential resource for researchers, laboratory workers, and policymakers. Some additional key points from the NIH Guidelines include:

- They apply to all research involving recombinant or synthetic nucleic acid molecules conducted in the United States, regardless of the source of funding.
- They are implemented through a system of institutional biosafety committees (IBCs). IBCs are responsible for reviewing and approving recombinant or synthetic nucleic acid molecule experiments, as well as for ensuring that the guidelines are being followed.
- They include a section on large-scale production of biological materials.
- They are subject to periodic review and update. The most recent update was issued in 2024.

The NIH Guidelines are a valuable resource for ensuring the safe and responsible conduct of research involving recombinant or synthetic nucleic acid molecules. However, the guidelines only cover research performed inside a laboratory setting and do not cover any field research or research involving release on synthetic biology experiments (such as gene drives or recombinant mosquitos) into an ecosystem. The responsibility for oversight is then transferred to other regulatory agencies, such as USDA or EPA. This fragmented approach of regulation and enforcement creates confusion and inconsistencies in understanding, interpretation, and implementation of best practices for Biorisk Management (BRM).

5.2.6 Biodefense in the Age of Synthetic Biology Report

The Biodefense in the Age of Synthetic Biology report is a comprehensive assessment of the potential security risks posed by synthetic biology.[125] The report was commissioned by the U.S. Department of Defense and was published by the National Academies of Sciences, Engineering, and Medicine in 2016.

The report identifies several potential biodefense vulnerabilities that could be exploited by malicious actors. These vulnerabilities include the ability to:

- Engineer organisms that are more resistant to antibiotics or vaccines.
- Create novel pathogens that are more deadly or transmissible.
- Modify existing pathogens to target specific populations.
- Use synthetic biology to produce toxins or other harmful substances.

The report also discusses several options for mitigating these biodefense risks. These options include:

- Increasing international cooperation on biosecurity.
- Strengthening biosecurity regulations.
- Developing new detection and response capabilities.
- Educating the public about the risks of synthetic biology.

The main conclusion of the report is that the biodefense risks posed by synthetic biology are real and should be taken seriously. However, the report also concludes that these risks are manageable with the right policies and investments. Some additional key points from the report include:

- The report defines synthetic biology as "the design and construction of new biological entities, components, or systems, or the redesign of existing biological systems, to achieve specific goals."
- The report identifies three main categories of biodefense vulnerabilities: (1) the ability to create new pathogens, (2) the ability to modify existing pathogens, and (3) the ability to produce harmful substances.
- The report discusses several options for mitigating biodefense risks, including international cooperation, strengthened biosecurity regulations, new detection and response capabilities, and public education.
- The report concludes that the biodefense risks posed by synthetic biology are real and should be taken seriously, but these risks are manageable with the right policies and investments.

5.2.7 Consideration on Efforts in Biosecurity within the U.S.

Collectively, the United States has a robust set of policies and guidelines that highlight the importance of stringent safety and security measures, responsible governance, ethical considerations, and international cooperation in navigating the potential risks of synthetic biology research. These documents lay the groundwork for robust safety measures, adherence to ethical principles, and developing a responsive and flexible system for biosafety and biosecurity implementation. They emphasize the need for the establishment and enforcement of strict standards for research conduct involving synthetic biology, grounded in the principles of beneficence, nonmaleficence, justice, and respect for autonomy. Moreover, these documents underscore the importance of collective international effort in mitigating potential security risks associated with synthetic biology, pointing toward the need for ongoing dialogue and cooperation at both the national and global levels. However, there are areas that are missed or not captured by these policies. For example, most of the regulations pertain to laboratories or research institutions that are performing government-funded research. On the other hand, if the research is privately funded, the researchers or organizations are not required to abide by these policies or guidelines.[126] Additionally, these policies and guidelines are not mandatory, often vague, and open to different interpretations, and enforcement is relegated to the institutions themselves through IBC committees and compliance offices. This leads to irregular understanding, implementation, and enforcement of guidelines nationwide. **This book advocates for adoption of unified standards, policies, and guidelines that are applicable and synchronized across different government agencies, capturing both publicly and privately funded research, as well as stringent unified enforcement through bodies external to the institutions being regulated.**

5.3 PERSPECTIVES ON BIOSECURITY OUTSIDE OF THE UNITED STATES

5.3.1 The Canadian Biosafety Standard

The Canadian Biosafety Standard (CBS), Third Edition is a national standard for facilities where regulated human and terrestrial animal pathogens and toxins are handled or stored.[127] The CBS was published in 2022 and replaced the second edition, which was published in 2015.

The CBS sets out the physical containment requirements, operational practice requirements, and performance and verification testing requirements that must be met to safely handle or store human and terrestrial animal pathogens and toxins. The CBS is based on a risk-based approach, and the requirements vary depending on the risk posed by the pathogens or toxins being handled.

The main conclusion of the CBS is that the safe and responsible handling of regulated pathogens and toxins requires a comprehensive biosafety program. The CBS provides a framework for developing and implementing such a program. Additionally:

- The CBS applies to all facilities in Canada that handle or store regulated pathogens and toxins, regardless of the size or type of facility.
- The CBS is a performance-based standard, which means that facilities have flexibility in meeting the requirements.
- The CBS is subject to periodic review and update.

The CBS is an essential resource for anyone involved in the safe and responsible handling of regulated pathogens and toxins in Canada. This document provides a comprehensive framework for developing and implementing a biosafety program, and it is an invaluable tool for protecting researchers, laboratory workers, the public, and the environment. In addition to the main conclusion, some other key takeaways from the CBS include:

- Biosafety is an essential component of any laboratory that works with regulated pathogens and toxins.

- The level of biosafety required depends on the risk posed by the pathogens or toxins being worked with.
- Biosafety programs should be tailored to the specific needs of the laboratory.
- Biosafety training should be provided to all laboratory personnel.
- Regulated pathogens and toxins should be handled and stored in a safe manner.
- Biosafety incidents should be reported and investigated promptly.

The CBS is a valuable resource for anyone in a laboratory that handles regulated pathogens and toxins. The CBS guides on a wide range of biosafety topics, and it is an essential tool for protecting researchers, laboratory workers, the public, and the environment.

Additionally, the Canadian Biosafety Handbook (2nd Ed), which is no longer in effect still contains valuable information.[128] They also produced the Public Health Agency of Canada (PHAC) Training Portal[129] and the Pathogens Safety Data Sheet.[130]

5.3.2 The European Biosafety Network and the European Commission's Directorate-General for Health and Food Safety

The European Union has a multifaceted approach to ensuring biosafety and biosecurity, focusing on healthcare and plant health. This section outlines the guidelines and recommendations provided by two prominent bodies within the EU: the European Biosafety Network (EBN)[131] and the European Commission's Directorate-General for Health and Food Safety.[132]

The European Biosafety Network (EBN), an assembly of organizations and individuals focused on promoting biosafety within the European Union, has published a comprehensive document outlining the crucial role of biosafety in the healthcare sector. The EBN actively provides an array of resources, such as training courses, toolkits, and publications, and organizes events and conferences on biosafety, offering a wide platform for knowledge sharing and engagement.

The EBN emphasizes that biosafety, which is critical for protecting healthcare workers and patients from exposure to hazardous substances, is a shared responsibility. It underlines the role of every healthcare worker in upholding biosafety practices within their respective environments.

The EBN advocates for the vital role of biosafety in the healthcare sector, emphasizing the shared responsibility among all healthcare workers. Through the outlined measures, the EBN provides a pragmatic guide to enhance biosafety practices in healthcare settings across the European Union. By consolidating these strategies, healthcare institutions can significantly bolster their commitment to ensuring a safe working environment, thereby protecting both their staff and the patients they serve.

The Directorate-General for Health and Food Safety of the European Commission provides a comprehensive overview of the importance of plant health and biosecurity within the European Union. It emphasizes the need for effective measures to protect crops, food sources, and the environment from potentially damaging pests and diseases. According to the document, the responsibility of maintaining plant health and biosecurity is shared by all stakeholders, thus emphasizing the collective effort required for the effective mitigation of biosecurity risks.

It also underscores the importance of partnerships with other countries and organizations to promote plant health and biosecurity further. Notably, the European Union continues to invest significantly in research and development to enhance plant health and biosecurity.

The guidelines set forth by the European Commission's Directorate-General for Health and Food Safety underscore the necessity of maintaining plant health and biosecurity within the European Union. The document provides a comprehensive framework of practical steps to bolster plant health and biosecurity within the region, underpinning a collective responsibility among all stakeholders in the face of evolving biosecurity risks.

5.3.3 African Union Biosafety and Biosecurity Initiative (AU-BBI)

The African Union (AU) is pioneering strides in biosafety and biosecurity through the African Union Biosafety and Biosecurity Initiative (AU-BBI),[133,134] launched in 2019. Designed as a comprehensive framework, the AU-BBI targets the enhancement of biosafety and biosecurity across AU member countries. The framework lays out several significant components:

- **Policy and Regulatory Frameworks**: A core aim of the AU-BBI is to develop and fortify biosafety and biosecurity policy and regulatory frameworks at a national level. This aims to ensure a standardized and comprehensive approach to managing biological threats across the continent.
- **Capacity Building**: The AU-BBI underlines the necessity to boost the capacity of AU nations to implement effective and appropriate biosafety and biosecurity measures. This includes fostering human resources, technological capabilities, and infrastructural support.
- **Monitoring and Evaluation**: The AU-BBI has integrated mechanisms to continuously monitor and evaluate the implementation of biosafety and biosecurity measures. This aims to ensure the effectiveness and accountability of biosecurity management across AU countries.

The AU-BBI, grounded in the principles of precaution and sustainable development, seeks to safeguard human health, environmental integrity, and security in Africa. This framework emerges as a critical initiative to tackle biosafety and biosecurity risks across the continent. The AU-BBI is a collaborative endeavor, bringing together the AU, the WHO, and other international organizations. The partnership aims to leverage collective expertise and resources to implement the framework effectively. This has materialized through regional and national projects under the AU-BBI, aimed at improving biosafety and biosecurity on the ground.

The AU-BBI is expected to significantly enhance the management of biosafety and biosecurity across the AU. The comprehensive and well-coordinated approach of the AU-BBI offers an effective and promising solution to the ever-growing challenges of biosafety and biosecurity in Africa.

5.3.4 Australia's Biosecurity Act

In the context of escalating global biosecurity risks such as bioterrorism and emerging infectious diseases, the Australian government enacted the Biosecurity Act 2015.[135,136] This pivotal piece of legislation aims to safeguard Australia from the biosecurity threats associated with biological agents and pests.

The Biosecurity Act 2015 provides a comprehensive and dynamic legal framework to regulate the importation, exportation, and movement of biological agents and pests within and across Australian borders. Designed to be adaptable to evolving biosecurity risks, the act applies universally to all biological agents and pests, irrespective of their origin or destination.

One of the primary objectives of the Act is preventing accidental or deliberate release of biological agents and pests. It enforces various measures such as rigorous import and export controls, stringent quarantine protocols, and active surveillance to achieve this objective.

In addition, the Biosecurity Act 2015 outlines comprehensive strategies to respond to biosecurity incidents. It prescribes the establishment of emergency response plans and the development of contingency plans to mitigate the impact of any biosecurity threats. This proactive approach allows for swift and effective responses to potential biosecurity incidents.

The Biosecurity Act 2015, therefore, serves as an essential tool in protecting Australia from biosecurity risks associated with biological agents and pests. Its enactment is a timely response to the growing global threat of bioterrorism and the emergence of infectious diseases.

The Biosecurity Act 2015 represents Australia's robust legislative response to biosecurity threats. By offering a comprehensive framework for regulating biological agents and pests and incorporating measures to prevent their release and respond to biosecurity incidents, the act is crucial in Australia's commitment to safeguarding national and global biosecurity. This framework's flexibility and adaptability ensure its ongoing relevance in the face of rapidly evolving biosecurity risks.

5.3.5 The Southeast Asia Strategic Biosecurity Dialogue

The Southeast Asia Strategic Biosecurity Dialogue (SEAS Dialogue) was established in 2014 as an initiative by the Johns Hopkins Center for Health Security and the University of the Philippines Manila.[137] Serving as a regional multilateral platform, the dialogue aims to promote engagement, collaboration, and exchange of knowledge on biosecurity within Southeast Asia.

The SEAS Dialogue offers a dynamic and flexible environment for addressing a broad array of biosecurity issues. It provides a unique forum where countries within the region can engage in meaningful dialogues, share best practices, and build trust. These interactions foster a cohesive regional approach to biosecurity, building a solid cooperation network among Southeast Asian nations.

The dialogue has been instrumental in raising awareness about biosecurity issues within the region, making it an invaluable tool for biosecurity education and advocacy.

By promoting knowledge sharing and collaboration, the SEAS Dialogue has played a significant role in fortifying biosecurity cooperation within Southeast Asia.

To date, the SEAS Dialogue has held five meetings, with each discussion fortifying the regional collaborative network. These dialogues have culminated in important outputs that underscore the initiative's impact. A notable achievement is the formulation of a declaration on biosecurity within Southeast Asia, which embodies the shared commitment of the participating nations to biosecurity. Furthermore, the dialogue has produced a roadmap for enhancing biosecurity cooperation in the region, providing strategic guidance for future initiatives.

The SEAS Dialogue exemplifies a successful regional initiative that has significantly contributed to biosecurity strengthening within Southeast Asia. Its flexible platform encourages open discussions, knowledge exchange, and collaboration. As a testament to its success, the SEAS Dialogue has facilitated critical outputs that further consolidate regional efforts toward biosecurity. As such, the SEAS Dialogue is a prime example of the power of multilateral dialogue in addressing global challenges such as biosecurity.

5.3.6 The Tianjin Biosecurity Guidelines for Codes of Conduct

The Tianjin Biosecurity Guidelines for Codes of Conduct for Scientists is a set of ten guiding principles and standards of conduct designed to promote responsible science practice and strengthen biosecurity governance at national and institutional levels in China.[138]

The guidelines were developed by an international group of experts in biosecurity and ethics, and they were endorsed by the InterAcademy Partnership in 2021. The guidelines are intended to be used by scientists, institutions, and policymakers to help ensure that bioscience research is conducted safely and responsibly. The ten guiding principles of the Tianjin Guidelines are:

1. Respect for human life and relevant social ethics.
2. Responsibility to use biosciences for peaceful purposes that benefit humankind.
3. Promotion of a culture of responsible conduct in biosciences.
4. Guarding against the misuse of science for malicious purposes.
5. Prevention of unintentional harm.
6. Openness and transparency.
7. Responsible sharing of information.
8. Education and training.
9. Institutional biosecurity measures.
10. International cooperation.

The main conclusion of the Tianjin Guidelines is that responsible science practice and strong biosecurity governance are essential to prevent the misuse of bioscience research for malicious purposes. The guidelines provide a set of principles and standards that can help ensure that bioscience research is conducted safely and responsibly. Some additional key points from the Tianjin Guidelines include:

- The guidelines focus on preventing intentional misuse of bioscience research, but they also emphasize the importance of preventing unintentional harm.
- The guidelines are designed to be flexible and adaptable to different national and institutional contexts.
- The guidelines are intended to be used with other biosecurity frameworks, such as BTWC.
- The Tianjin Guidelines are an important contribution to the global effort to promote responsible science practice and strengthen biosecurity governance. The guidelines provide a valuable resource for scientists, institutions, and policymakers working to ensure that bioscience research is conducted safely and responsibly.

5.3.7 Guidelines for Biosafety and Biosecurity in the Middle East and North Africa Region

North Africa and the Middle East present a unique biosafety and biosecurity landscape, owing to their proximity to conflict zones, diverse ecosystems, and varying stages of development in scientific infrastructure. A broad-based analysis of biosafety and biosecurity in these regions emerges from three critical documents that offer comprehensive insights into the existing challenges and potential solutions.

"Biological Security Priorities in the Middle East" by the United Nations Interregional Crime and Justice Research Institute (UNICRI) underlines the region's heightened vulnerability to biosafety and biosecurity risks.[139] The gaps in biosafety and biosecurity are substantial, with a pressing need for enhanced cooperation and coordination in the region. Recommendations include strengthening national biosafety and biosecurity frameworks, enhancing capacity-building initiatives, and fostering deeper collaboration on regional biosafety and biosecurity issues.

"Biosafety Initiatives in BMENA Region: Identification of Gaps and Advances," a report delving into the biosecurity implications of synthetic biology, flags the double-edged nature of this rapidly developing field.[140] While synthetic biology holds great promise for advancing biotechnology, it also engenders significant biosecurity risks. This necessitates robust biosecurity measures to curtail the potential misuse of synthetic biology for malicious ends. The report urges the development of strong biosecurity policies and regulations, capacity-building in biosecurity, fostering international cooperation, and elevating public and policymaker awareness of synthetic biology's biosecurity risks.

Lastly, the research article titled "Reasons for and Barriers to Biosafety and Biosecurity Training in Health-Related Organizations in Africa, Middle East, and Central Asia: Findings from GIBACHT Training Needs Assessments 2018–2019" echoes these concerns in Africa.[141] It highlights the need for strong biosafety and biosecurity measures to mitigate risks associated with synthetic biology. It underlines the importance of creating robust biosafety and biosecurity policies and regulations, fostering capacity building, promoting international cooperation, and raising awareness about the risks of synthetic biology among the public and policymakers in Africa.

Collectively, these documents underscore the pressing need to address biosafety and biosecurity issues in North Africa and the Middle East, particularly concerning synthetic biology. There is a clear consensus on the need to develop strong regulatory frameworks, enhance capacity building, and cultivate international cooperation to foster a robust biosafety and biosecurity environment in these regions. In the face of real and serious biosecurity risks, these regions must weigh the promise of synthetic biology against its potential perils. As such, a balanced approach that recognizes both the potential benefits and the inherent risks is vital for effective biosafety and biosecurity in North Africa and the Middle East.

5.3.8 The Brazilian Biosafety Network

In the ever-evolving landscape of Synthetic Biology, Brazil's emphasis on biosafety is a noteworthy model to explore. The Brazilian Biosafety Network (SB3) has published an illuminating document that outlines the critical role of biosafety in protecting not just human health, but also the well-being of animals and the broader environment from the threats posed by harmful biological agents.[142]

The SB3 document underscores that biosafety is a shared responsibility that transcends institutional or organizational boundaries. Every stakeholder, from research scientists to policymakers, plays a crucial role in maintaining biosafety. The document further explores a spectrum of practical measures for enhancing biosafety in Brazil. These strategies include:

- **Adoption and Implementation of National Biosafety Frameworks**: To ensure a unified and coherent approach to biosafety, a national-level policy or framework is essential. This includes protocols for safe lab practices, containment procedures, and response plans for biosafety incidents.
- **Strengthening Biosafety Capacity in Laboratories**: This includes implementing robust safety protocols, investing in safe and secure equipment and facilities, and regularly reviewing and updating safety practices.
- **Improving Biosafety Awareness and Practices**: Educational initiatives can play a significant role in raising awareness about the importance of biosafety and fostering a culture of safety in every aspect of biological research and application.
- **Increasing Collaboration and Coordination on Biosafety**: Fostering collaboration between different stakeholders can lead to shared learning, improved standards, and a cohesive response to biosafety challenges.
- **Investing in Research and Development on Biosafety**: Ongoing research is vital to keep pace with new challenges and developments in synthetic biology.

The SB3 is a nexus for organizations and individuals committed to promoting biosafety and biosecurity within Brazil. To facilitate its mission, the network offers a range of resources including training courses, toolkits, and publications tailored to biosafety. In addition, SB3 organizes events and conferences to foster a broader discourse on biosafety issues, challenges, and solutions.

5.3.9 Indian Council of Agricultural Research Guidelines & Agricultural Biosecurity Bill

India has established a comprehensive legal framework for biosafety and biosecurity to regulate the safe development and use of biotechnology. These regulations are continually updated to reflect scientific advancements and to address emerging biosafety risks. The Indian Council of Agricultural Research (ICAR) discusses several key laws and regulations in this domain[143]:

- **The Environment (Protection) Act, 1986**: As the foundation for environmental protection in India, this Act mandates obtaining prior environmental clearance for activities involving biological agents. It supports the principle of precaution and fosters sustainable development.
- **The Biological Diversity Act, 2002**: This Act was instituted for the conservation of biological diversity, sustainable use of its components, and equitable sharing of benefits arising from the utilization of biological resources. Prior approval from the National Biodiversity Authority is required for activities involving biological resources, further underlining India's commitment to biosafety.
- **The Rules for the Manufacture, Use, Import, Export, and Storage of Hazardous Microorganisms, Genetically Engineered Organisms or Cells, 1989**: These rules present comprehensive requirements for handling, storing, and disposing hazardous microorganisms and genetically engineered organisms or cells.
- **The Guidelines for the Safety Assessment of Foods Derived from Genetically Engineered Plants, 2005**: This document guides the safety assessment of foods derived from genetically engineered plants, thus ensuring public health safety.

These laws and regulations underscore India's comprehensive approach to BRM. The foundational principles of precaution and sustainable development guide these regulations, ensuring the protection of human health, the environment, and the conservation of biological diversity.

As scientific advancements continue to evolve and new biosafety risks emerge, the Indian government remains committed to updating these laws and regulations. In this way, India ensures that biotechnology develops responsibly and sustainably, contributing to global efforts in managing potential risks associated with biological research, including synthetic biology. The legal framework is a helpful guide for practical steps in ensuring the safe development and use of biotechnology in India.

Moreover, the Agricultural Biosecurity Bill, introduced in India in 2013, is significant legislation designed to establish the Agricultural Biosecurity Authority of India (ABAI).[144] The ABAI would be crucial in regulating the import and export of plants, animals, and related products in the country.

The primary responsibilities of the ABAI, as outlined in the bill, include the prevention of the introduction and spread of pests and diseases in India. The ABAI would thus play a key role in ensuring that imported plants and animals are free from harmful pests and diseases. Moreover, the Authority would oversee the devising and implementation of biosecurity measures to protect India's vital agricultural sector.

This bill is of considerable importance in bolstering the defenses of India's agricultural sector against pests and diseases, which can have detrimental effects on crop yield, livestock health, and the overall economic prosperity of the sector. Furthermore, it would contribute significantly to ensuring that India's agricultural exports meet international safety and quality standards.

At the time of introduction, the Agricultural Biosecurity Bill, 2013, was referred to the Lok Sabha, the lower house of Parliament. The bill has since been forwarded to the Standing Committee on Agriculture for a thorough examination. The committee's report, eagerly anticipated, is expected to shed further light on the practicalities and implications of the bill.

The Agricultural Biosecurity Bill, 2013, presents a comprehensive approach to safeguarding India's agricultural sector. It underscores the need for robust biosecurity practices to ensure that the nation's agricultural systems remain resilient in the face of potential biological threats. This legislation, if enacted, would provide significant guidelines and steps for enhancing biosecurity in India's agricultural sector, thereby making an essential contribution to the country's broader BRM.

5.3.10 Japan's Guidelines for Biosafety and Biosecurity in Laboratories

Japan has proactively established robust biosafety and biosecurity guidelines for laboratories in line with international standards. The Ministry of Education, Culture, Sports, Science, and Technology (MEXT) published "The Guidelines for Biosafety and Biosecurity in Laboratories" in 2020, a comprehensive document aligning with the International Health Regulations (2005) and the WHO Laboratory Biosafety Manual, Fourth Edition.[145]

The guidelines encompass a wide range of critical areas, including the classification of biological agents, the design and operation of laboratories, handling and disposal of biological agents, training and education, and incident response. The guidelines serve not only to ensure compliance with international standards within Japanese laboratories, but also as a valuable resource for laboratories worldwide. By providing practical guidance in implementing biosafety and biosecurity measures, they contribute to the global effort to manage the potential risks associated with biological research, including the emerging field of synthetic biology.

5.3.11 The Korean Virtual Biosecurity Center Guidelines

The Korean Virtual Biosecurity Center provides a comprehensive set of guidelines and insights into South Korea's biosafety and biosecurity framework.[146] This framework is built upon several key laws and regulations that aim to manage the complex challenges posed by the rapidly developing field of biotechnology, including synthetic biology. Key aspects of South Korea's biosafety and biosecurity legal and regulatory framework include:

- **The Biosafety Act**: This act serves as the cornerstone of biosafety in South Korea, governing the classification, handling, import, and export of biological agents. It stipulates the safety measures necessary for laboratories working with such agents, aiming to protect both lab workers and the broader environment.
- **The Biosecurity Act**: Designed to safeguard South Korea's national security, this act focuses on protecting critical infrastructures, preventing bioterrorism, and managing responses to biosecurity incidents. This is especially pertinent given the potential for misuse of synthetic biology technologies.
- **The National Biosafety and Biosecurity Plan**: This strategic plan outlines the government's approach to biosafety and biosecurity. It sets goals, objectives, and specific measures to be implemented, ensuring a holistic, proactive approach to managing the risks associated with biotechnology.

These documents collectively create a robust legal and regulatory framework responsive to scientific advancements and emerging biosecurity risks. In accordance with principles of precaution and sustainable development, this framework seeks to protect human health, the environment, and national security, while facilitating safe biotechnological development and use. The guidelines and regulations set by the Korean Virtual Biosecurity Center underscore the importance of a comprehensive and dynamic biosafety and biosecurity framework. They serve as a blueprint for managing the risks and opportunities associated with the use of synthetic biology, both for South Korea and for other nations seeking to navigate the complexities of this rapidly advancing field.

5.3.12 The Biosecurity Regime in Denmark

Denmark adopted a unique Biosecurity Regime (here regime is defined as a system or planned way of doing things) worth highlighting here. It is based on information presented at the BTWC meeting in Geneva on December 10, 2013.[147] Central to establishing a biosecurity regime in Denmark constitutes an overview of biohazardous locations and an umbrella structure for pre-existing legal frameworks. This regime also provides guidelines for organizing, supervising, and embedding biosecurity within organizations, along with methods for transmitting critical information to first responders. The regime further underscores the roles of various ministries in biosecurity management, thus accentuating the interdepartmental nature of the issue. Additionally, it introduces self-assessment toolkits – the CBRN-security toolkit and the Biosecurity toolkit. Both aim to augment biosecurity awareness, with the latter developed explicitly by an expert panel for assessing potential gaps in an organization's biosecurity regime. The Biosecurity toolkit also recommends good practices to enhance biosecurity levels.

One of the key aspects highlighted in the Biosecurity toolkit is the "eight priority areas of biosecurity." These priority areas encompass Physical Security, Personal Security, Material Control and Accountability, Transport Security, Information Security, Awareness, Response, and Organization Management. To ascertain an organization's biosecurity readiness, the toolkit's self-scan module poses a series of questions under each of these themes.

The Danish Biosecurity Regime underscores the significance of a systematic and comprehensive approach to biosecurity. Their Biosecurity toolkit is a practical tool for organizations to identify gaps in their biosecurity measures and bolster their biosecurity awareness and preparedness. No country in the world has stricter biosecurity legislation and enforcement than Denmark. They also produced a manual on how you build up a biosecurity system in a country, which is very accessible and does not have a counterpart anywhere else in the world.[148]

5.3.13 Argentina's Guidelines and Regulations for Biosafety and Biosecurity

The guidelines and regulations pertaining to biosafety and biosecurity in Argentina are dispersed across multiple sources, which cater to the diverse and complex needs of the country's biological sector.

Key among these sources are the resolutions issued by various governmental entities. For instance, Resolution 284/2013 from the National Commission for the Conservation and Sustainable Use of Biodiversity (CONICET) stipulates the infrastructural, procedural, and personnel training prerequisites for laboratories working with biological agents belonging to risk groups 1 and 2.[149]

The National Directorate of Biosecurity (ANLIS-Malbrán), under Resolution 1029/2019, imposes requirements on labs dealing with biological agents of higher risk, namely groups 3 and 4.[150] This includes mandatory protocols for the physical infrastructure, safety procedures, and staff training.

The Ministry of Agriculture, Livestock and Fisheries extended these requirements further through Resolution 179/2022.[151] This legislation encompasses all biological risk groups, from 1 through 4, ensuring a comprehensive coverage of safety standards in labs handling these agents.

These resolutions, anchored in international standards such as the International Health Regulations (2005) and the WHO Laboratory Biosafety Manual, 4th Edition, provide robust guidelines for the secure handling, storage, and disposal of biological agents within laboratory settings.

To supplement these legislative resources, additional guidance and support are available through the Argentine Biosafety Network (Red Argentina de Bioseguridad). This network serves as a critical hub for information and resources pertinent to BRM in Argentina.

The WHO Biosafety and Biosecurity Training Programme for Latin America and the Caribbean offers specialized training on biosafety and biosecurity to professionals in the region. This program reinforces the existing regulatory frameworks and contributes to developing a proficient workforce adept at managing biological agents securely and effectively. This holistic approach serves to enhance the biosafety and biosecurity landscape in Argentina.

5.3.14 Consideration on Efforts in Biosecurity Outside of the U.S.

The standards and regulations set by different regions worldwide highlight the global nature of BRM in synthetic biology. They showcase a comprehensive approach to securing bioscience labs and ensuring the safe handling of biological materials. Each standard offers a unique perspective shaped by the region's cultural, economic, and environmental context, yet all share common themes of risk mitigation, standardization of safety protocols, and the promotion of responsible scientific practices.

The Canadian Biosafety Standard (CBS) and other global standards, such as the European Biosafety Network and the European Commission's Directorate-General for Health and Food Safety, share a common trait in promoting a culture of safety and responsibility in handling regulated pathogens and toxins. These standards provide a framework for developing comprehensive biosafety programs applicable to various facilities, regardless of size or type.

Standards in emerging economies and developing nations, such as the African Union Biosafety and Biosecurity Initiative (AU-BBI) and the Indian Council of Agricultural Research Guidelines & Agricultural Biosecurity Bill, highlight the importance of establishing robust biosafety and biosecurity frameworks at the national level. They underscore the need for capacity building and continuous monitoring and evaluation of implemented measures to manage biological threats effectively.

Australia's Biosecurity Act, Southeast Asia Strategic Biosecurity Dialogue, and Japan's Guidelines for Biosafety and Biosecurity in Laboratories showcase the power of legislation and multilateral dialogue in managing biosecurity risks. These initiatives reflect the necessity for dynamic legal frameworks adaptable to evolving biosecurity risks and the importance of open discussions, knowledge exchange, and collaboration among regional stakeholders.

Regulations in the Middle East and North Africa Region and Latin America demonstrate the importance of regional context in shaping BRM. These regions face unique challenges due to their proximity to conflict zones, diverse ecosystems, and varying stages of development in scientific infrastructure, underscoring the need for strong regional cooperation and coordination in BRM.

Finally, the Tianjin Biosecurity Guidelines for Codes of Conduct in China and the Korean Virtual Biosecurity Center Guidelines emphasize responsible scientific practices and strong biosecurity governance. They underscore the role of scientists, institutions, and policymakers in ensuring biosafety and biosecurity and offer a comprehensive set of guiding principles.

Each of these standards and regulations underscores the need for a comprehensive, tailored approach to biosafety and biosecurity in synthetic biology, while underscoring the importance of global cooperation, capacity building, risk mitigation, and promoting responsible scientific practices for safer and more secure bioscience laboratories worldwide.

5.4 GLOBAL HEALTH SECURITY

Biosecurity is a critical component of global health security. As the world becomes increasingly interconnected through travel and trade, the risk of dangerous pathogens crossing borders and threatening public health continues to grow. Recent disease outbreaks, such as the COVID-19 pandemic, have demonstrated the profound health, economic, and social impacts of inadequate BRM measures. When developing policies, guidelines, rules, regulations, and other frameworks for global health security, it is essential that biorisk management be considered. Implementation and enforcement of BRM measures mitigate bioterrorism and pandemic threats. By incorporating biosecurity into the global health security agenda, nations can work collectively to prevent catastrophic biological events, detect threats early, and respond rapidly and effectively when outbreaks do occur. A comprehensive approach to global health security is not possible without addressing biosecurity.

5.5 CONCLUDING THOUGHTS

Creating harmonized global regulatory frameworks in synthetic biology is timely and crucial as rapid advancements often outpace current regulatory measures, leading to potential gaps in biosecurity standards. A strong call for international collaboration in developing dynamic and adaptive regulatory frameworks is an important step forward. These frameworks must be comprehensive, addressing current challenges and anticipating future developments to ensure synthetic biology's safe and ethical progression. There is a diversity of unharmonized regulatory frameworks and policies governing synthetic biology across different regions. The development of dynamic, robust, and adaptive regulatory frameworks is essential. Such frameworks should prioritize safety, uphold ethical integrity, gain societal acceptance and support, and balance innovation with responsibility. Concerted international efforts are vital to bridging regulatory gaps and facilitating responsible innovation and collaboration in this rapidly evolving field. This comprehensive approach is essential to guide the field toward a future where innovation is advanced but also responsible and inclusive.

5.6 KEY TAKEAWAYS

- **Diverse Regulatory Landscapes**: Acknowledge the varying stages of regulatory development in different regions.
- **Need for Harmonization**: Highlights the challenge of harmonizing regulations internationally.
- **Dynamic Regulatory Frameworks**: Emphasize the need for robust, comprehensive, yet flexible and adaptive frameworks.
- **Ethical and Safety Concerns**: Focus on addressing ethical integrity and safety in synthetic biology.
- **International Collaboration**: Calls for global cooperation to bridge regulatory gaps and promote responsible innovation.
- **Biosecurity**: Stresses the importance of international perspectives on biosecurity in synthetic biology.

5.7 THOUGHT-PROVOKING QUESTIONS

5.1 How effective has the Biological and Toxin Weapons Convention (BTWC) been in mitigating biosecurity risks posed by synthetic biology in various world regions? What are the gaps still missing in its implementation? While 185 out of 193 countries have signed and ratified this document, nine countries have not.

5.2 How effective was the role of the Cartagena Protocol on Biosafety in regulating the applications of synthetic biology in different regions of the world? What are the gaps still missing in its implementation? While 185 out of 193 countries have signed and ratified this document, nine countries have not.

5.3 What is missing from the current international documents for biorisk management to ensure:
   i. Long-term international cooperation in building a robust regulatory framework for synthetic biology?
   ii. Regulatory frameworks that ensure equitable access and benefits of synthetic biology across different societies?
   iii. Regulatory policies that influence the trajectory of synthetic biology research and its applications in different fields?

5.4 What strategies are needed for improving transparency and information sharing in synthetic biology research without compromising proprietary information?

5.5 How can differing national regulations on synthetic biology be harmonized to be integrated into a global, consistent set of standards? What challenges can be expected in harmonizing synthetic biology regulations across countries and regions?

6 Regulatory, Ethical, and Policy Considerations for Synthetic Biology

The complex landscape of regulatory, ethical, and policy considerations is integral to synthetic biology. This chapter builds on the previous chapter, emphasizing the need to establish robust regulatory frameworks to oversee the rapidly evolving realm of synthetic biology. Here, the focus is on the ethical considerations inherent in synthetic biology, including the ethical responsibilities of scientists in this field, the need for transparency, public engagement, and consideration of long-term impacts. It emphasizes the importance of an integrated approach that combines regulatory measures, ethical principles, and sound policymaking. Such an approach is crucial for harnessing the benefits of synthetic biology while minimizing its risks and making the benefits equitable globally. The chapter advocates for ongoing dialogue among scientists, policymakers, ethicists, and the public to navigate synthetic biology's complex ethical and regulatory landscape, ensuring responsible and sustainable advancement in the field.

DOI: 10.1201/9781003423171-6

6.1 CURRENT REGULATORY MEASURES FOR SYNTHETIC BIOLOGY

Regulatory measures related to synthetic biology are complex and vary widely from country to country. They also depend on the specific applications of the technology (e.g., medical, agricultural, and environmental). An overview of some common areas of regulation and the general regulatory landscape:

- **Research Oversight**: Many countries have regulations and guidelines that govern research on genetically modified organisms (GMOs), which typically also cover organisms engineered through synthetic biology. These may include requirements for containment and safety measures, risk assessments, and oversight by institutional or national bioethics committees.
- **Product Approval**: In many jurisdictions, synthetic biology products must go through an approval process before they can be commercialized. This typically involves demonstrating the safety and efficacy of the product. The specific requirements can vary widely depending on the nature of the product and the jurisdiction.
- **Environmental Release**: The release of genetically engineered organisms into the environment is subject to stringent regulation. This usually involves a detailed risk assessment and may also require monitoring after the release by the World Health Organization and/or the World Organization for Animal Health.
- **Dual-Use Research of Concern (DURC)**: Many countries have specific regulations or guidelines related to DURC, which is research that, while intended to benefit society, could be misused to harm public health or the environment.
- **Gene Drive**: Gene drive, a technology that can spread a specific set of genes throughout a population, is an area of intense regulatory interest due to its potential to cause rapid, irreversible changes to ecosystems. Currently, there is an ongoing debate about how this technology should be regulated.
- **Human Genome Editing**: Countries differ widely in their regulatory approach to human genome editing. Some, like the United Kingdom, allow certain types of research under strict regulation, while others, like the United States, have more restrictive policies, particularly on germline editing (which can be inherited or passed on from one generation to the next).

The regulation of synthetic biology is evolving along with the field itself. As technology advances, it will be necessary to continually reassess and update regulatory measures to ensure they are effective, proportionate, and able to address new challenges and opportunities.

6.2 ETHICAL GUIDELINES AND INSTITUTIONAL OVERSIGHT

Ethical guidelines and institutional oversight play key roles in ensuring responsible research and application of synthetic biology. A few aspects where these principles are applied include:

- **Institutional Biorisk Committees (IBCs)**: In many countries, research involving recombinant or synthetic nucleic acid molecules is overseen by IBCs at the institutional level. These committees, which typically include scientists, safety experts, and members of the local community, review and approve proposed research projects to ensure they comply with safety and ethical standards.
- **Ethics Review Boards**: Projects that involve human subjects, such as clinical trials of gene therapies, must be reviewed by ethics committees or institutional review boards (IRBs). These bodies review research protocols to ensure that they protect the rights and welfare of the research subjects, including informed consent and privacy protections.
- **Dual-Use Research of Concern (DURC)**: DURC is the research that, while intended to be beneficial, could be misused to threaten public health, agriculture, or the environment. Guidelines have been established to identify DURC and implement risk mitigation measures. Oversight bodies often include representatives from various governmental departments to ensure a comprehensive review.
- **Genetic Engineering and Genome Editing Guidelines**: Various organizations have published guidelines for genetic engineering and genome editing research. For example, the National Academies of Sciences, Engineering, and Medicine in the United States have issued reports guiding human genome editing. The World Health Organization is also developing global standards. For further information, see Sections 5.1, 5.1.5, and 5.3.
- **Professional Codes of Conduct**: Many scientific societies and professional organizations have codes of conduct that their members are expected to adhere to. These often include ethical principles relevant to synthetic biology, such as honesty, integrity, respect for life and the environment, and social responsibility.

These measures together aim to ensure that synthetic biology is conducted in a manner that is safe, ethical, and socially responsible. However, given the rapid pace of advancements in this field, ongoing dialogue and iterative revisiting of these regulations and ethical guidelines is crucial to address emerging challenges and considerations appropriately at a much faster pace.

6.3 OTHER ETHICAL CONSIDERATIONS

Delving into synthetic biology entails a journey into a myriad of ethical quandaries. From creating novel life forms to applying gene-editing technologies to humans, synthetic biology opens possibilities that stretch the moral fabric of society.

One such concern is the idea of human germline editing. The capacity to manipulate the genetic makeup of human embryos presents a host of ethical issues. While it holds the potential to eliminate hereditary diseases, it also gives rise to fears of "designer babies" and the possibility of creating a socioeconomic divide between those who can afford such technologies and those who cannot. Moreover, there is the concern of unintended genetic modifications that could cause unanticipated harm, or the modification of traits whose functions are not fully understood, potentially leading to unforeseen health and societal implications.

Synthetic biology also allows for creation of entirely new organisms, which brings us face-to-face with complex ethical questions. What rights would such organisms have, if any? What ecological impact could they have if they inadvertently escaped into the wild? Would creating synthetic organisms infringe on notions of the sanctity of life? The potential for unintended consequences regarding ecological disruption or potential harm to existing species necessitates careful consideration and robust regulation.

Another significant ethical concern is the commodification of life. As synthetic biology allows us to engineer life forms for specific purposes, this could lead to a situation where organisms are viewed purely as commodities. This stance could devalue the intrinsic worth of living organisms and disrupt our relationship with the natural world.

Ethical considerations in synthetic biology, therefore, need to be addressed through an inclusive dialogue that involves scientists, ethicists, policymakers, and the public. Ethical discussions take time to develop and reach conclusions over long periods of time; however, we are facing a technology that changes overnight. If one country or region declares a "stand-down" until risks have been understood thoroughly, this does not necessarily mean that other countries will follow suit. Thus, we need to work together to ensure that the development and application of synthetic biology is guided by ethical principles that uphold respect for life and biodiversity, social justice, and the well-being of all. Furthermore, it is crucial to foster transparency in synthetic biology research and applications to build public trust and to ensure the public can make informed decisions about these rapidly developing technologies.

6.4 EQUITABLE ACCESS TO BIOSECURITY RESOURCES

Equity in biosecurity refers to ensuring fair and equal access to biosecurity measures and protections across different regions, institutions, and socioeconomic backgrounds. It recognizes that access to resources and capabilities needed to address biosecurity risks can be disproportionately distributed, with low-resource settings facing greater challenges. This issue of equity in biosecurity deserves more attention to address the potential disparities and promote a more inclusive and resilient global biosecurity landscape. Here are some key points to consider:

- **Resource Disparities**: Low-resource settings, such as developing countries or underfunded institutions, may lack the necessary financial, technological, and human resources to implement robust biosecurity measures. This can include limited access to trained personnel, quality infrastructure, appropriate safety equipment, or up-to-date technologies. Addressing these resource disparities is crucial to ensuring that all regions have an equal opportunity to protect against and respond to biosecurity threats.
- **Capacity Building and Knowledge Transfer**: Efforts should focus on building the capacity and expertise of individuals and institutions in low-resource settings. This can involve providing training programs, knowledge-sharing platforms, and mentorship opportunities to enhance their understanding of biosecurity principles and practices. Collaborative

partnerships between high-resource and low-resource settings can facilitate knowledge transfer and foster a more equitable distribution of expertise.

- **International Cooperation and Funding**: International cooperation is vital for addressing equity in biosecurity. Developed countries and international organizations should prioritize supporting capacity-building initiatives in low-resource settings through funding, technical assistance, and knowledge exchange. This includes facilitating access to affordable and reliable biosecurity technologies, promoting technology transfer, and supporting the development and implementation of biosecurity regulations in underserved regions.

By addressing equity in biosecurity, we can strive toward a more inclusive and globally connected approach to biosecurity. This requires recognizing and addressing the disparities in resources, knowledge, and access to ensure that all regions and communities can effectively protect themselves and contribute to global biosecurity efforts.

6.5 ENVIRONMENTAL IMPACT OF SYNTHETIC BIOLOGY

As synthetic biology propels us forward into a new era of genetic manipulation, it is important that we tread carefully, especially when considering the release of GMOs into the environment. This is because these GMOs could have significant ecological impacts.

For instance, let us consider an engineered strain of a crop plant with enhanced resistance to pests. The intentional release of this GMO could initially lead to increased crop yields, reducing the need for pesticide use and contributing to food security. However, it could also disrupt local ecosystems in unforeseen ways. The engineered crops could outcompete native plants, leading to a decrease in biodiversity. Furthermore, the pests targeted by the crop's enhanced resistance could decrease in population, affecting other species that rely on these pests as a food source. This could potentially set off a chain reaction of changes in the ecosystem.

Another example is the creation of genetically engineered mosquitoes designed to combat diseases such as malaria or dengue. These mosquitoes, modified to either be sterile or carry a lethal gene, are intentionally released into the environment with the intent of reducing the population of disease-carrying mosquitoes. While this sounds promising from a public health perspective, there are potential ecological concerns. The reduction or elimination of the mosquito population could have cascading effects on the ecosystem, as other organisms that rely on mosquitoes for food might be adversely affected. Moreover, the sudden void in the ecosystem could be filled by another, potentially more harmful, species.

Therefore, while synthetic biology promises innovative solutions to many global challenges, it is crucial to consider potential environmental impacts. Any intentional release of GMOs into the environment should be preceded by rigorous risk assessments to anticipate potential ecological impacts. Strict regulatory oversight is required to ensure the safety of these interventions. Also, continuous monitoring post-release is essential to track the long-term impacts and to enable timely responses to any adverse effects that might emerge. This balanced approach can help harness the benefits of synthetic biology while safeguarding our ecosystems.

6.6 INTELLECTUAL PROPERTY RIGHTS AND ACCESS TO TECHNOLOGIES

Intellectual property (IP) rights play a significant role in synthetic biology, as they do in other areas of biotechnology. Some key aspects to consider include:

6.6.1 Patents

IP rights, particularly patents, play a fundamental role in synthetic biology. They offer the creators of new organisms, genetic constructs, methods, or applications exclusive rights to their inventions for a defined period, typically 20 years. This exclusivity can fuel innovation, offering researchers and corporations a financial incentive to invest time, effort, and resources into developing new synthetic biology technologies.

However, the exclusivity afforded by patents can also create challenges. When key methods, components, or organisms are patented, other researchers or organizations may be barred from using them unless they obtain a license from the patent holder. Licensing terms may not always be reasonable or accessible, potentially hindering research and limiting the progress of synthetic biology.

To illustrate, consider a company that develops a novel gene synthesis method. If the company patents this method and refuses to license it (or sets prohibitively expensive licensing fees), other researchers may be prevented from using it to develop their innovations. This could slow the pace of research and development in synthetic biology.

Balancing this need for protection against IP theft with the desire to build a global, collaborative community is a complex task. One potential solution could be establishing open-source platforms or frameworks in synthetic biology. In such models, researchers could willingly share their findings, methods, and genetic sequences, promoting a more collaborative and less restricted research environment.

Another approach could involve patent pools, where multiple patents are grouped and made available for licensing under predetermined terms. This approach could help address the issue of "patent thickets," where a dense web of overlapping patents makes it difficult for researchers to navigate the IP landscape.

Moreover, certain fair and reasonable licensing practices could also be encouraged. For instance, researchers or institutions could commit to licensing their patented technology on Fair, Reasonable, and Non-Discriminatory (FRAND) terms.

The goal should be to foster an environment where innovation is rewarded and protected, but not at the expense of scientific collaboration and progress. Given the global nature of many challenges, synthetic biology seeks to address – from pandemics to climate change – the value of building a collaborative, global community in synthetic biology cannot be overstated.

6.6.2 Access and Benefit-Sharing

The equitable sharing of benefits and access to genetic resources is an essential aspect of synthetic biology. The Nagoya Protocol, a supplementary agreement to the Convention on Biological Diversity (CBD), plays a crucial role, providing a framework to ensure that benefits from using genetic resources are shared fairly and equitably.

With synthetic biology, the use of genetic resources extends beyond physical materials, encompassing digital sequence information as well. This digital information – such as DNA sequence data – can be freely shared and used to design and create synthetic organisms without requiring physical access to the original genetic resources. While this opens immense opportunities for research and innovation, it also creates challenges for the traditional understanding of access and benefit-sharing under the Nagoya Protocol.

However, the ease of sharing digital sequence information can risk bypassing the access and benefit-sharing provisions of the Nagoya Protocol. For instance, a researcher might obtain a DNA sequence from an online database, originating from a plant in a specific country, and use that sequence to engineer a bacterium to produce a valuable compound. In this scenario, the benefits are not being shared with the country providing the genetic resource, which contradicts the spirit of the Nagoya Protocol.

Addressing this issue will require updates to international legal and regulatory frameworks and the development of new mechanisms to track the use of digital sequence information and ensure that benefits are shared equitably. This might involve innovative solutions, such as blockchain technology to track the use of genetic resources, or new models for benefit-sharing in the digital age.

Moreover, developing guidelines or agreements for data sharing in synthetic biology research could contribute to a more equitable international scientific community. These could ensure that researchers in low-resource settings have access to the same datasets as those in high-resource settings, helping to democratize access to information and avoid the concentration of knowledge and resources.

While the digitization of genetic resources presents challenges for traditional access and benefit-sharing frameworks, it also offers opportunities to rethink these frameworks in ways that promote both innovation and equity in synthetic biology.

6.6.3 Open Science and Biohacking

There are movements within the synthetic biology community toward more open science, including open access to genetic parts (like the BioBricks Foundation) and open-source hardware for biotechnology. However, these approaches can clash with traditional IP rights. In addition, the growing popularity of biohacking or DIY biology raises questions about access to technology and the potential for misuse.

6.6.4 Proprietary Technology and Access to Medicine

When synthetic biology is used to produce medical products, such as drugs or vaccines, IP rights can have a direct impact on access to these medicines. High prices driven by monopolies can make it difficult for patients, particularly in low- and middle-income countries, to access these treatments. On the other hand, strong IP protection can incentivize companies to invest in expensive and risky research and development.

6.6.5 Ethics of Patenting Life

The ethics of patenting life forms, an issue that becomes increasingly pertinent with the advancement of synthetic biology, draws much debate. At the heart of the discussion are concerns about the commodification of life, accessibility of genetic resources, and potential impacts on biodiversity and Indigenous communities.

Patenting in synthetic biology can involve granting exclusive rights over specific sequences of DNA, methods of gene editing, GMOs, or entirely synthetic life forms. On the one hand, patents can provide necessary incentives for innovation by offering a temporary monopoly to inventors and allowing them to recoup the investment made in research and development. This has been instrumental in driving forward biotechnological breakthroughs that have far-reaching societal benefits.

On the other hand, the concept of owning life, even in its synthetic form, can be morally troubling to some. Life, in its natural form, is considered by many to be a common heritage of humankind and not something that should be privately owned. Synthetic biology complicates this issue by blurring the lines between what is considered natural and what is synthetic.

Further, there are concerns about access to patented life forms or their genetic sequences. This could create a divide where resources are only available to those who can afford them, potentially limiting the scope of research and the development of new applications, particularly in low-resource settings.

One notable example is the case of the first synthetic bacterium, *Mycoplasma laboratorium*. Created by the J. Craig Venter Institute in 2010, the bacterium had an entirely synthetic genome, and it was patented by the Institute. The move sparked controversy, with critics arguing that it sets a dangerous precedent for the ownership of synthetic life and could have profound implications for biosecurity, biodiversity, and the sharing of genetic resources.

Given these complexities, managing IP rights in synthetic biology requires a careful balance. It is crucial to incentivize innovation and reward the investment of time, effort, and resources. Simultaneously, it is important to ensure broad access to technologies, maintain the free exchange of ideas, and consider the potential social and economic impacts.

6.7 RECOMMENDATIONS FOR POLICY, RESEARCH, AND EDUCATION

As synthetic biology continues to advance, there is a need for ongoing work in policy, research, and education to ensure that this technology is developed and used safely and responsibly. Some recommendations in each of these areas include the following.

6.7.1 Policy

- **Develop Clear Regulations**: Governments should establish clear regulations for synthetic biology research and applications. These regulations should balance the need for safety and oversight with the need to encourage innovation.
- **Establish International Standards**: The international and transboundary nature of synthetic biology research and its potential applications underscore

the necessity of establishing global standards. A unified, international approach to biorisk management can help facilitate effective collaboration and information sharing across borders, ensuring a consistent level of safety and security, while addressing ethical, legal, and social implications.

One way to achieve this is by developing a global synthetic biology biosecurity manual. This manual could provide comprehensive guidelines for the containment, handling, and transportation of genetically engineered organisms. The contents could include guidance on physical security, personnel safety, information security, bioethical considerations, and risk assessment methodologies that are tailored to the unique risks posed by synthetic biology research.

This global manual could draw from and harmonize existing national and regional biosafety and biosecurity regulations, standards set by organizations like the World Health Organization (WHO), and best practices from industry and academia. It could also account for different resource settings, providing adaptable solutions for synthetic biology labs in low-resource contexts.

Regular updates and revisions would be necessary to keep pace with the rapid advancements in the field. These could be facilitated by a governing body that includes representatives from diverse geographical, institutional, and professional backgrounds.

For example, the International Gene Synthesis Consortium (IGSC), an association of gene synthesis companies and organizations, has developed a Harmonized Screening Protocol. The protocol provides a standard for screening DNA sequence orders to identify and manage sequences of concern, which could be used in biological weapons or to create harmful biological agents. While not a comprehensive biosecurity manual, this is an example of how international organizations can come together to establish standards in response to the specific risks posed by synthetic biology.

- **Incorporate Stakeholder Input**: Policymakers should seek input from a wide range of stakeholders, including scientists, ethicists, industry representatives, and members of the public when developing regulations and policies.

6.7.2 Research

- **Prioritize Safety Research**: There should be ongoing research into the safety and biosecurity aspects of synthetic biology. This should include studying the behavior of synthetic organisms in different environments, developing methods for containing and monitoring synthetic organisms, and exploring potential misuse scenarios.
- **Promote Responsible Innovation**: Researchers should strive to incorporate safety and ethical considerations into their work from the very beginning. This could be achieved through practices like safety-by-design and ethical-by-design approaches.
- **Support Interdisciplinary Collaboration**: Given the complex nature of synthetic biology, collaboration between researchers from different disciplines (such as biology, engineering, ethics, and social science) is essential.

6.7.3 Education

- **Integrate Ethics and Safety into Education**: Education in synthetic biology should not only focus on the technical aspects of the field, but also on the ethical, social, and safety implications. This will help to prepare the next generation of synthetic biologists to navigate these issues.
- **Promote Public Understanding**: There should be efforts to improve public understanding of synthetic biology, such as through science communication initiatives, public lectures, and museum exhibits. This can help to foster informed public debate about the direction of the field.
- **Develop Professional Standards**: Professional organizations can play a role in developing codes of conduct or ethical guidelines for synthetic biologists. These can guide professionals in the field and help to establish norms of responsible behavior.

By addressing these recommendations, we can help to ensure that synthetic biology develops in a way that is safe, ethical, and beneficial for society.

6.8 CONCLUDING THOUGHTS

The advancement of synthetic biology opens new scientific frontiers and challenges our understanding and appreciation of life, which requires approaching advancements with ethical integrity and social responsibility. As we delve deeper into synthetic biology, it challenges us to redefine our ethical boundaries and responsibilities. These challenges call for a concerted, inclusive dialogue involving diverse perspectives to navigate these uncharted ethical territories. The ethical implications of manipulating life must be conscientiously confronted, from creating new life forms to editing human genes. Scientists, biorisk management professionals, and policymakers bear the moral responsibility to consider the long-term impacts of their work, not just on human health and disease but also on societal norms, environmental integrity, and the essence of life itself. The ethical landscape of synthetic biology extends beyond the laboratory, implicating broader societal values and raising questions about the sanctity of life, equity, and the commodification of living organisms. Therefore, there is an imperative for an ethical compass to guide the journey of synthetic biology, ensuring that its profound powers are harnessed with respect for life, social justice, and the collective well-being of our planet.

6.9 KEY TAKEAWAYS

- **Ethical Guidelines**: Stress the importance of ethical guidelines and institutional oversight in synthetic biology research.
- **Equitable Access**: Focuses on the need for equitable access to biosecurity resources and technologies.
- **Environmental Impact**: Discusses the potential ecological effects of synthetic biology, emphasizing the need for risk assessments and regulatory oversight.

- **Intellectual Property Rights**: Address the role and challenges of IP rights in synthetic biology, balancing innovation with access.
- **Policy Research and Education**: Recommends ongoing work in policy research and education to guide the safe and responsible development of synthetic biology.

6.10 THOUGHT-PROVOKING QUESTIONS

6.1 What are the top five disparities in resources and capabilities between high-resource and low-resource settings that must be addressed?

6.2 How can the current strategies for equity in biosecurity be updated so that low-resource settings can practice research in synthetic biology safely, and what is a timeline to achieve this leveling of the playing field?

6.3 What is still missing in a rigorous risk assessment plan before any GMO release to ensure strict regulatory oversight, continuous post-release monitoring, and a quick-response mechanism to deal with adverse effects? What regulatory body should be responsible for developing and maintaining a comprehensive plan to minimize the potential environmental impact of synthetic biology, especially regarding the release of GMOs into the environment?

6.4 What innovative solution could be proposed to address the challenges associated with access and benefit-sharing provisions of the Nagoya Protocol in the era of digital sequence information?

6.5 What would it take to implement blockchain technology to track the use of genetic resources, ensuring benefits are shared equitably?

6.6 How can "open science" and biohacking movements be reconciled with traditional IP rights? How can a balanced approach be designed and implemented to involve creating open-source platforms for sharing findings and methods alongside reasonable IP protection to incentivize innovation?

6.7 What should be included in an ideal global synthetic biology biosecurity manual? What key components should it have, and how should it be updated over time?

6.8 How can policy development ensure the responsible application of synthetic biology in various sectors?

6.9 How can potential risks in synthetic biology be effectively managed without stifling innovation?

6.10 How can stakeholders effectively engage in discussions around the ethical, regulatory, and policy considerations for synthetic biology?

6.11 How can we ensure that synthetic biology's benefits are distributed equitably across different populations and regions?

7 Biorisk Management in Synthetic Biology

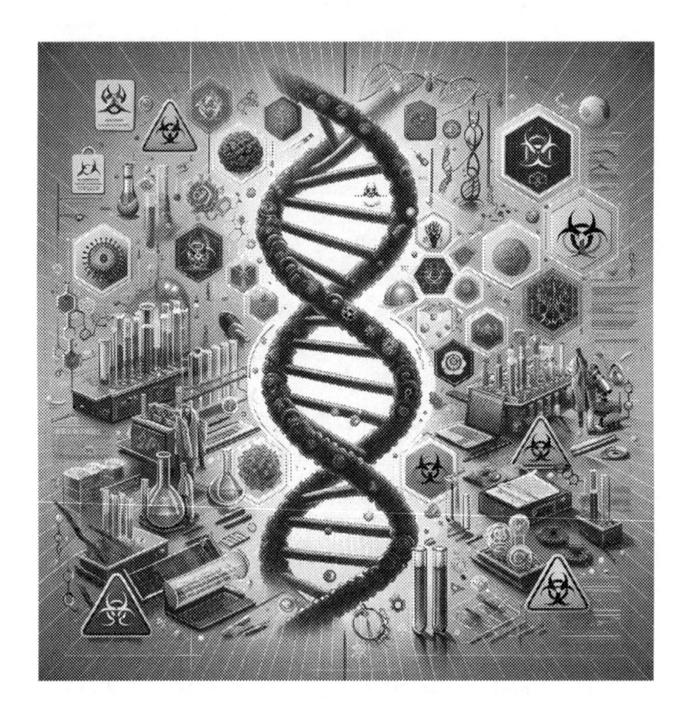

This chapter explores the considerations, strategies, and practices related to biosafety and biosecurity in the rapidly advancing field of synthetic biology. It refers to the analysis and mitigation of risks associated with the use and handling of biological agents and toxins in laboratories and industries, aiming to protect health, safety, and the environment. With the advent of synthetic biology and the potential it holds for healthcare, energy, and environmental sectors, the importance of having a robust biorisk management system cannot be overstated. National and institutional biorisk management policies and programs are crucial to ensure that the benefits of synthetic biology are harnessed safely and responsibly. They can provide a regulatory framework to guide the development and implementation of safety measures, fostering an environment of security, trust, and accountability for researchers, industry professionals, and the public. This chapter delves into the importance of biorisk management in synthetic biology, exploring existing policies and protocols, highlighting the need for standardized international practices, and underscoring the ethical

DOI: 10.1201/9781003423171-7

considerations associated with these efforts. It provides a comprehensive understanding of the interplay between synthetic biology and biorisk management, thereby informing the creation of effective policies and safety protocols that can keep pace with the fast-evolving scientific landscape.

7.1 WHO'S "BIORISK MANAGEMENT – LABORATORY BIOSECURITY GUIDANCE"

The World Health Organization's (WHO) 2006 publication,[152] "Biorisk Management – Laboratory Biosecurity Guidance," is a guideline for countries and institutions handling valuable biological materials. The document outlines strategies for identifying, preventing, and mitigating biosecurity risks. Recommendations for laboratory biosecurity include routine biorisk evaluations, outlining responsibilities for personnel, developing a culture of awareness, ethics, and respect, and advancing policies that enable seamless scientific data and material exchanges without obstructing research. Further recommendations include bolstering ties between scientific, technical, and security realms, providing pertinent training to laboratory staff, and augmenting emergency response strategies, acknowledging that biorisk management, though effective, cannot invariably neutralize every conceivable threat.

This document, published in 2006, is outdated and does not address emerging biosecurity risks associated with fields like synthetic biology, artificial intelligence, or cyberbiosecurity. There is an urgent need for an updated version, especially considering the COVID-19 pandemic. WHO member states need more relevant guidance on managing VBM and biosecurity risks. The WHO experts have been working on this update and the community is eagerly waiting for this updated version to be released soon.

7.2 CEN WORKSHOP AGREEMENT (CWA 15793)

The CEN[a] Workshop Agreement (CWA) 15793 represents a vital set of voluntary guidelines central to biorisk management within laboratory settings.[27] Introduced by the European Committee for Standardization in 2011, it details a comprehensive set of requirements for the handling and storing of biological agents and toxins, thereby playing an instrumental role in biosafety and biosecurity.

CWA 15793 is a tool designed to assist organizations in implementing a robust biorisk management system. This agreement is constructed on the foundations of the International Health Regulations (2005) and the WHO's Laboratory Biosafety Manual (Fourth Edition). It encompasses not just the risk to employees operating within laboratory environments, but also the broader community and the environment.

The overarching aim of CWA 15793 is to enable organizations dealing with biological agents and toxins to mitigate the potential risks associated with these activities effectively. Developed by a working group of experts from CEN, CWA 15793's applicability extends beyond the borders of Europe. The agreement is designed for global

[a] CEN: Comité Européen de Normalisation (French) or European Committee for Standardization (English)

use, complementing other national and international standards on biosafety and biosecurity. Thus, it provides a global standard for the safe and secure handling and storage of biological agents and toxins in laboratories. Although it is expired, it is freely available and therefore still has value as a resource for the worldwide community.

Despite its widespread usage, the CWA 15793 standard has notable limitations. Its primary constraint is that it is a voluntary standard, thus lacking enforceability. Furthermore, the CWA 15793 focuses on the generic risk management of biological agents, irrespective of their classification or containment levels. While this standard has recently been superseded by the new ISO35001 standard, the rapid advancements in synthetic biology since the publication of CWA 15793 have not been adequately captured. The pace and scope of these developments call for updated guidelines, highlighting a pressing need for refreshed and comprehensive guidance.

7.3 ISO 35001:2019 BIORISK MANAGEMENT FOR LABORATORIES AND OTHER RELATED ORGANIZATIONS

The International Organization for Standardization (ISO) developed and published the ISO 35001:2019 standard,[153] which provides a biorisk management system that is procedures oriented pertinent to the handling, storage, and transportation of biological agents and toxins. It was developed by an ISO technical committee composed of global experts from the field. It is meant to replace the CWA15793 CEN agreement. However, it is essential to note that ISO 35001:2019 does not cover activities involving the use of biological agents and toxins in clinical trials, military applications, or acts of bioterrorism. It relies on third-party validation and certification, which can be costly for low-resource countries and organizations; it is more focused on biosafety than biosecurity; and it is not currently mandatory by any governments or industries.

ISO 35001:2019 outlines several requirements for an effective biorisk management system. These include leadership and commitment demonstrated by top management, effective planning, and implementation of biorisk management activities, and constant monitoring, measurement, analysis, and evaluation of the system's performance. It outlines how to manage any changes to the biorisk management system effectively and continually strive for improvement. Organizations that adhere to the ISO 35001:2019 standard will develop a process-oriented biorisk management program for handling biological agents and toxins, significantly reducing their risk of accidents and incidents, increasing regulatory compliance, improving public confidence, and an enhanced organizational reputation. Implementing ISO 35001:2019 can aid organizations in mitigating these risks. This standard offers a framework that guides the safe and responsible conduct of research and development in synthetic biology. Organizations can identify, assess, and control the risks associated with the use of biological agents and toxins in synthetic biology, enabling regulatory compliance.

Implementation of ISO 35001:2019, while beneficial, also presents some challenges. Notably, the cost barrier can be substantial, especially for low-resource countries and organizations. Implementing this standard requires significant investment in personnel training, equipment, infrastructure, and ongoing maintenance to ensure compliance. The cost of ISO certification itself can be prohibitive. For organizations in low-resource settings, these costs may make the standard inaccessible, potentially limiting their ability to engage in biological research safely and responsibly. Therefore, although ISO 35001:2019 is an important tool for promoting biosecurity, its application requires careful consideration of resource availability and specific local context.

The lack of accessibility to the ISO 35001:2019 standard for some organizations and countries, combined with the lack of enforceability of the CWA 15793, leaves the biosecurity field with a significant gap in unified, harmonized, and accessible standards that all agree to abide by locally and globally. In synthetic biology, this poses an important vulnerability as experts and stakeholders are left to navigate novel waters without enforceable guidance and guardrails.

7.4 ARTIFICIAL INTELLIGENCE-AIDED METHODS FOR ASSESSING BIOSECURITY RISKS

I propose that a biosecurity risk assessment be carried out before any synthetic biology project is developed or approved. Biosecurity risk assessment in synthetic biology is an important process to ensure the safe development and application of new technologies while minimizing the potential for misuse or unintended consequences. This process will be described in detail in Section 11.1. Here, I outline the main steps in the process while highlighting the opportunities for AI to improve the process:

- **Hazard and Risk Identification**: The first step in biosecurity risk management is identifying potential risks, ranging from accidental release of engineered organisms to deliberate misuse for harmful purposes. This step often requires a comprehensive review of the technology, its potential applications, and the potential pathways for harm. Here, AI can play a transformative role by automating the process of risk identification. Advanced AI algorithms can analyze vast amounts of data to spot potential hazards that might go unnoticed by human analysts, especially in the complexity of synthetic biology. Machine learning models can be trained to recognize patterns associated with risk factors, thereby enhancing the reliability of risk identification.
- **Risk Analysis**: Once potential hazards have been identified, they need to be analyzed to understand their likelihood and consequences. While qualitative methods rely on expert opinion, quantitative methods typically involve mathematical modeling or statistical analysis. AI has shown tremendous promise in improving the accuracy and speed of semiquantitative risk analysis. AI algorithms can process large datasets quickly, spot trends, and make predictions, thereby providing more accurate estimates of risk probability and potential impact. Machine learning techniques, such as neural networks or decision tree analysis, can simulate and analyze complex risk scenarios that would be challenging for traditional methods.
- **Risk Evaluation**: Following risk analysis, risks are evaluated to determine acceptable ones and those that require mitigation measures. Typically, this involves a comparison of the risks with the potential benefits of the technology, as well as a consideration of societal values and risk tolerance. AI can significantly enhance this process, particularly when assessing complex and multifaceted risks. Through data-driven approaches, AI can help develop

sophisticated models that weigh risks against benefits, considering likelihood, consequences, and potential for harm. This can lead to more robust, objective, and transparent evaluations, enabling safer and more responsible development and deployment of synthetic biology technologies.

- **Risk Management**: For risks deemed unacceptable, risk management measures are developed. These can range from modifying laboratory procedures to reduce accidental release risks to establishing regulations or oversight mechanisms to prevent misuse. AI can significantly assist in this process. Machine learning algorithms can sift through massive amounts of data to identify potential risks, while predictive modeling can anticipate future risk scenarios. AI can also help monitor compliance with risk management measures, instantly identifying and alerting to any breaches or deviations.
- **Risk Communication**: Clear, transparent communication about risks with all stakeholders, including researchers, policymakers, and the public, is an essential part of biosecurity. AI can augment this process by personalizing risk communication for different audiences. It can analyze comprehension levels and adapt the communication style, enhancing understanding and engagement. AI chatbots and virtual assistants can also provide real-time answers to questions, further improving the communication process.
- **Review and Update**: Given the rapid pace of advancement in synthetic biology, risk assessments must be frequently reviewed and updated as new information becomes available. AI can automate this process, continuously scanning the latest research, news, and regulatory updates for relevant information. AI algorithms can then use this data to update risk assessments in real time, ensuring that they always reflect the current state of knowledge and risk; while a biosecurity expert helps curate and confirm that the AI updates are accurate and correct.

While these steps provide a basic framework for biosecurity risk assessment, it is important to note that this is a complex process that requires a deep understanding of both the technical aspects of synthetic biology and the broader societal context. Although AI technology can help automate and augment this process, it will not replace the roles and responsibilities of experts. It also requires a multidisciplinary approach, involving experts in molecular biology, ethics, law, and risk assessment.

7.5 STRATEGIES FOR CONTAINMENT AND PREVENTION

Containment and prevention are crucial elements of managing the potential risks associated with synthetic biology. Strategies can vary depending on the context, but some key strategies often employed include:

- **Physical Containment**: This refers to measures that prevent the escape of organisms from a lab setting. These measures include containment facilities and equipment, such as specialized ventilation systems, autoclaves for sterilizing waste, and biosafety cabinets for handling organisms.

- **Biological Containment**: This involves engineering organisms so that they cannot survive outside of the lab. This could involve making organisms dependent on a nutrient that is not available in the natural environment, or engineering "kill switches" that cause organisms to self-destruct under certain conditions.
- **"Safe by Design" Concept**: "Safe by Design" in synthetic biology is a proactive approach to minimize risks associated with the use and release of engineered organisms. For enhanced biosecurity, safety measures should be integrated during the design phase in any synthetic biology context. This should cover a range of things from engineering organisms and planning projects to the building of new laboratories. By predicting potential issues beforehand, preventive action can be taken to eliminate risks. Some ways "Safe by Design" can be implemented include:
  - **Auxotrophy**: Engineering organisms to be auxotrophic, or unable to synthesize a specific critical compound, is a common approach. This means the organism cannot survive without being supplied with that specific compound, which is typically not available in the wild. If such an organism were to escape, it would not be able to survive or reproduce.
  - **Genetic Safeguards**: Another strategy is to build in genetic safeguards such as "kill switches" that cause the organism to self-destruct under certain conditions or "suicide genes" that trigger cell death in response to specific stimuli. These mechanisms can help to ensure that an organism does not persist in an outside environment if it were to escape from the laboratory.
  - **Biological Containment**: Biological containment refers to the use of host-vector systems that prevent the survival of recombinant DNA in non-laboratory environments. This involves using genetically engineered vectors that can only replicate in specific host strains.
  - **Degradable Genetic Material**: One recent innovation is xeno nucleic acids (XNAs), synthetic alternatives to DNA that can carry genetic information but degrade quickly in natural environments, making horizontal gene transfer unlikely.
  - **Geocontainment**: Certain organisms can be engineered to survive only in specific geographic areas (geocontainment). This could involve engineering an organism to require a specific nutrient only found in the target area or to be unable to survive the climatic conditions outside the target area.
  - **Safe Laboratory Practices**: The "Safe by Design" approach can also apply to laboratory design and practices. This includes physical containment measures such as sealed laboratories, air filtration systems, and specialized equipment to ensure organisms cannot escape.
  - **Regulatory Compliance**: Ensuring regulatory compliance in the design of synthetic organisms is another aspect of the "Safe by Design" approach. This can involve conducting a detailed risk assessment during the design phase to ensure that the organism complies with all local and international biosecurity regulations.

Adopting a "Safe by Design" approach can lead to safer experiments, minimize the chances of harmful events, and foster a safety culture in synthetic biology. Embedding safety considerations into the design process, not only reduces the risks associated with synthetic biology but also increases public trust in this rapidly advancing field.

- **Procedural Containment**: These are protocols and procedures designed to reduce the risk of accidental release, such as guidelines for handling and disposing of organisms, emergency response plans for accidents, and training for staff on biosecurity procedures.
- **Regulatory Measures**: These include laws and regulations that govern the use of synthetic biology, such as requirements for risk assessments, approvals for certain activities, and inspections of facilities.
- **Ethical Guidelines and Oversight**: This involves professional codes of conduct, ethics review boards, and other mechanisms to ensure ethical behavior by researchers.
- **Public Engagement and Transparency**: Engaging the public and maintaining transparency about synthetic biology activities can help build trust, promote responsible behavior, and enable early identification of potential issues.
- **International Cooperation and Capacity Building**: Because synthetic biology is a global endeavor, international cooperation is crucial for effective biosecurity. This can include sharing best practices, building capacity in countries with less experience in synthetic biology, and developing international norms and agreements.
- **Ongoing Research and Innovation**: Ongoing research into new containment and prevention strategies is essential, given the rapid pace of advancement in synthetic biology.

These strategies can help ensure that synthetic biology is conducted responsibly while minimizing the risks of accidental release or misuse of engineered organisms.

7.6 BIOSAFETY PROTOCOLS AND BEST PRACTICES

Biosafety is a key aspect of synthetic biology, aimed at protecting people and the environment from potential harm. Some of the best practices and protocols commonly used in the field include:

- **Risk Assessment**: Before beginning any experiment or project, it is crucial to perform a risk assessment to identify potential hazards and decide on appropriate safety measures.
- **Lab Safety Training**: All researchers and staff should undergo regular safety training, which includes general lab safety as well as specific practices for working with genetically modified organisms or potentially hazardous substances.
- **Use of Personal Protective Equipment (PPE)**: Lab coats, gloves, safety glasses, and other PPE should be used as appropriate, depending on the level of risk.

- **Physical Containment Measures**: These include using containment facilities and equipment, such as biosafety cabinets, autoclaves, and specialized ventilation systems.
- **Biological Containment Measures**: Strategies might include engineering organisms to be auxotrophic (requiring a specific nutrient not found in the environment to survive) or incorporating "kill switches" into organisms that trigger self-destruction under specific conditions.
- **Waste Management**: All waste should be decontaminated before disposal. This typically involves autoclaving or chemical disinfection for biological waste and following local regulations for the disposal of chemicals.
- **Emergency Response Procedures**: Labs should have clear plans and procedures for responding to emergencies, such as spills, accidents, or exposures.
- **Regular Audits and Inspections**: Regular safety audits can help identify potential issues before they become serious problems.
- **Documentation and Record Keeping**: Keeping accurate records of experiments, risk assessments, safety measures, and incidents can help in learning from past experiences and improving safety practices.
- **Regular Review and Update of Protocols**: Biosafety protocols should be regularly reviewed and updated to consider new information, changes in regulations, or advances in technology.

Remember, biosafety practices can vary depending on the type of work being done and the potential risks involved. Always refer to local regulations and guidelines, as well as international best practices, when developing biosafety protocols for synthetic biology.

7.7 EMERGENCY PREPAREDNESS AND RESPONSE

Emergency preparedness and response are crucial components of biorisk management in synthetic biology. Emergency management focuses on the management of disasters, which are events larger than a community can handle on its own, such as floods, earthquakes, hurricanes, or terrorist attacks.[154] Disaster preparedness requires a combination of coordination from individuals, households, organizations, local government all the way to higher levels of government or international response. Emergency preparedness and response management can be categorized into preparedness, response, mitigation, and recovery, although other terms such as disaster risk reduction and prevention are also common. The outcome of emergency preparedness and response is to mitigate the harmful impact of disasters on the community. It involves planning and putting in place protocols to handle potential emergencies that could stem from accidents, misuse, or unforeseen impacts of the technology. Here is how it can be approached:

- **Emergency Planning**: The first step in emergency planning is to understand the broad spectrum of potential risks and emergencies that could arise from the research or application of synthetic biology. This process should cover a broad range of potential threats – not just biological hazards like the accidental release of synthetic organisms, but also security threats such as data breaches, insider threats, or cyberattacks that could result in the theft of sensitive data or intellectual property.

Insider threats can pose significant challenges as they come from individuals who have authorized access to laboratory facilities or information systems, making detection and prevention more complex. Protocols must be in place to monitor unusual behavior and measures for immediate suspension of access.

Cyber threats, particularly hacking, pose unique challenges in our digital age. Hackers could potentially infiltrate network systems, manipulate data, disrupt operations, or even gain control of laboratory equipment. The emergency response plan must include robust cybersecurity measures, regular system updates and checks, and a rapid response protocol for suspected breaches.

Infrastructure threats, such as power loss or network downages, also need consideration. Sudden power loss could lead to equipment failure, loss of critical samples or data, or containment breaches. Strategies such as backup power supplies, off-site data backups, and redundant network systems can help mitigate these risks.

Once these potential emergencies have been identified and comprehensively understood, detailed response plans should be developed. These plans should clearly outline the steps to be taken in the event of such emergencies, detailing communication chains, specific roles and responsibilities, evacuation plans if necessary, and procedures for containment and mitigation. Regular training and drills can ensure that all involved personnel know their roles and can respond efficiently when an emergency arises.

- **Roles and Responsibilities**: Each person involved in the project should have a clear understanding of their roles and responsibilities in an emergency. This includes not only frontline responders, but also decision-makers who will coordinate the response and communicate with stakeholders.
- **Training and Drills**: Regular training sessions and drills should be conducted to ensure everyone knows what to do in an emergency. This helps to reduce panic and confusion, allowing the situation to be managed more efficiently.
- **Communication Plan**: A clear communication plan should be in place, including protocols for notifying relevant internal and external stakeholders (such as public health agencies, law enforcement, or potentially affected communities) and providing regular updates as the situation evolves.
- **Review and Revision**: After an emergency, a thorough review should be conducted to identify what went well and what could be improved. The emergency response plan should then be revised based on these insights. Even without an emergency, plans should be periodically reviewed and updated to reflect changes in the research environment or new insights into potential risks.

Emergency preparedness and response are an integral part of a proactive and robust biorisk management strategy in synthetic biology, helping to mitigate the impact of emergencies and ensure a swift and effective response when they occur.

7.8 BIOSECURITY INCIDENT MANAGEMENT

Biosecurity incident management is different from emergency preparedness and response in that it specifically addresses the management of an incident such as an intentional biological spill, intentional release, or an act to injure a laboratorian. It is critical to include biosecurity incident management plans, which deal with responding effectively and quickly to accidents, spills, releases, intentional theft, hacking, or other potential risks associated with synthetic biology. Key components of such a strategy include:

- **Incident Response Planning**: Preparation is key for managing any emergency. This includes creating and maintaining detailed response plans for different types of incidents, including the accidental release of a synthetic organism, hacking of the computer systems, intentional release for nefarious purposes, or personnel exposure (accidental or intentional).
- **Designated Response Personnel**: Assign specific roles and responsibilities for incident response. This typically involves designating personnel who will take the lead in managing the response and coordinating with external agencies as necessary.
- **Training and Drills**: Regular training and drills ensure that all staff know what to do in an emergency. This training should be tailored to the specific risks associated with the lab's synthetic biology work and should include everyone, from senior researchers to junior staff.
- **Notification Procedures**: Have clear procedures for notifying relevant authorities, such as institutional safety officers, local health departments, or regulatory agencies. Also, set up a communication plan to keep all lab members and potentially affected parties informed about the situation.
- **Containment and Cleanup**: When applicable, include protocols for containing the incident and cleaning up afterward. This may involve specialized equipment or procedures for handling synthetic organisms or other biological hazards.
- **Health Monitoring**: In case of potential exposure to hazardous substances or organisms, arrange for appropriate health monitoring of affected individuals. This could involve medical examinations, lab tests, or even post-exposure prophylaxis in some cases.
- **Incident Investigation and Follow-Up**: After the immediate incident is resolved, conduct a thorough investigation to understand the cause and how it can be prevented. Follow-up actions might involve changes to procedures, additional training, or physical modifications to the lab.
- **Documentation and Reporting**: Document all aspects of the incident and the response, and report as required to internal and external authorities. This documentation can help in reviewing and improving incident response procedures, and it is often a legal requirement.
- **Post-Incident Surveillance and Monitoring**: After a biosecurity incident, rigorous surveillance and monitoring procedures are crucial to ensure the effectiveness of the response and remediation efforts. This phase involves

the collection and analysis of relevant data, such as environmental sampling, epidemiological studies, or network activity logs, depending on the nature of the incident.

For instance, in the case of an accidental release of a genetically engineered organism, this could involve environmental monitoring to track the spread and survival of the organism, assessing potential interactions with native species, and studying any unexpected impacts on the local ecosystem. It might also necessitate health monitoring of potentially exposed individuals.

In a data breach or cyberattack, post-incident monitoring might involve analyzing network activity logs, identifying any compromised accounts or systems, and assessing the extent of data loss or corruption. Regular system audits and intrusion detection systems can help identify any ongoing or recurrent threats.

Post-incident surveillance and monitoring can provide valuable insights into the effectiveness of the response, identify areas for improvement, and guide the development of more robust measures for preventing future incidents. Any lessons learned should be incorporated into updated biosecurity policies and training programs, contributing to the continuous improvement of the overall biosecurity framework.

An effective incident response strategy will minimize the impact of any incident, protect the safety and security of lab personnel and the public, and ensure compliance with biosecurity regulations. It also helps to maintain public trust and confidence in synthetic biology.

7.9 INSTITUTIONAL BIORISK COMMITTEES (IBCs) & OVERSIGHT

In the dynamic landscape of biological research, where advancements in technology continuously reshape the potential risks and challenges, there is a growing need to continuously evaluate and enhance our approach to biosafety and biosecurity oversight. One critical step in this direction is my proposal to transform Institutional Biosafety Committees (IBCs) into Institutional Biorisk Committees (IBCs). This change is not merely semantic, but it reflects a more comprehensive approach to managing risks in the biosphere. While biosafety focuses on protecting individuals and the environment from exposure to biological agents, biosecurity extends this scope to encompass the prevention of loss, theft, misuse, diversion, or intentional release of these agents. By adopting the concept of Institutional Biorisk Committees, institutions can ensure a more holistic and integrated approach to managing biosafety and biosecurity, addressing the entire spectrum of risks associated with biological research and fortifying our collective defense against potential biological threats.

7.9.1 Institutional Biorisk Committees (IBCs)

I propose that Institutional Biosafety Committees be renamed Institutional Biorisk Committees (IBCs) to better capture their critical role in managing biorisk, rather than focusing solely on biosafety. IBCs, as discussed earlier (see Section 6.2), are key components of the biorisk management infrastructure for research involving recombinant or synthetic nucleic acid molecules, genetically modified organisms, and other potentially hazardous biological agents. IBCs are tasked with assessing and managing the risks associated with such research, and they play an essential role in ensuring biosecurity, safety, and compliance with regulations and guidelines. Some of the specific roles of IBCs include:

- **Approval of Research**: Before research involving potentially hazardous biological materials can commence, it needs to be reviewed and approved by the IBC. The committee will consider factors like the risks posed by the research, the qualifications of the research team, and the adequacy of the containment and safety measures.
- **Oversight and Monitoring**: IBCs are responsible for ongoing oversight of research activities. This can involve inspecting laboratories, monitoring compliance with safety procedures, and responding to any incidents or concerns.
- **Risk Assessment**: One of the primary responsibilities of IBCs is to review and approve risk assessments for the proposed research. This involves evaluating the potential risks posed by the research, such as the possibility of accidental release of genetically modified organisms, and the adequacy of proposed safety measures.
- **Development of Policies and Procedures**: IBCs often play a role in developing institutional policies and procedures related to biorisk management. This can involve guidelines for the safe handling of biological materials, emergency response plans, and training requirements.
- **Training and Education**: IBCs often oversee or contribute to biorisk management training and education for researchers. This ensures that those working with potentially hazardous biological materials are aware of the risks and know how to work safely.
- **Liaison with External Entities**: IBCs often act as a liaison between the institution and external entities such as regulatory agencies, the local community, or funding bodies. This can involve reporting on the institution's biorisk management practices, responding to audits or inspections, and addressing any concerns from the community.

By carrying out these functions, IBCs play a crucial role in managing biorisks, protecting the safety of researchers and the public, and ensuring research is conducted responsibly and in accordance with relevant regulations and guidelines.

7.9.2 Importance of Internal Review Process

The internal review process is a fundamental component of the biorisk management system, and a critical part of the functions conducted by IBCs. It serves as the primary mechanism by which research projects involving biohazardous materials are evaluated for potential risks to researchers, the public, and the environment, as well as the appropriateness of safety measures. Reasons why the internal review process is so crucial include:

- **Ensuring Compliance with Regulations**: The review process checks that the proposed research complies with applicable regulations and guidelines at local, national, and international levels. This includes ensuring that the research has the necessary permits or licenses and adheres to standards for biorisk management.
- **Protecting Researchers and the Public**: By evaluating and mitigating risks, the review process helps protect the safety of researchers working with hazardous biological materials and prevents harm to the broader public and the environment.
- **Quality Assurance**: The internal review process provides a mechanism for quality assurance, ensuring that research involving biohazardous materials is conducted responsibly, ethically, and to a high scientific standard.
- **Promoting Transparency and Accountability**: The review process promotes transparency and accountability in research by requiring that potential risks and safety measures be documented and reviewed by a committee.
- **Building Public Trust**: Rigorous internal review processes can help build public trust in the institution's research by demonstrating that potential risks are taken seriously and managed effectively.

For these reasons, the internal review process is a cornerstone of effective biorisk management in synthetic biology and a key function of IBCs. It is an ongoing process that requires continuous updates as research evolves and new information becomes available, ensuring that biosecurity measures keep pace with scientific progress.

7.10 BIOSECURITY LEVELS: DEFINITION AND DESCRIPTION

Laboratories should apply the relevant risk mitigation measures for the work being performed, as defined by organizations such as the Centers for Disease Control and Prevention (CDC) and the WHO. In the United States, there are four Biosafety Levels (BSLs), BSL-1 being the lowest and BSL-4 being the highest level of containment. Internationally, the community has moved to a more risk informed approach, with WHO defining "core requirements," "heightened control measures," and "maximum containment measures." Here, I propose a similar approach be defined and implemented for biosecurity levels.

Building upon the existing structure of Biosafety Levels (BSLs), I propose the introduction of Biosecurity Levels (BSec-Ls) to provide a tiered framework for biosecurity risk management in synthetic biology laboratories. Just like BSLs in the United States, BSec-Ls would be determined based on the potential biosecurity risk associated with the activities being performed in the lab, taking into account factors such as the nature of the biological agents being handled, the processes and techniques being used, and the potential for misuse or unauthorized access, as summarized in Table 7.1.

- **Biosecurity Level 1 (BSec-L1)**: This is the basic level of biosecurity that should be implemented in all synthetic biology labs and facilities. It includes access control, training personnel on biosecurity risks and procedures, inventory controls for biological materials, and basic physical security of labs. This aims to promote a general culture of biosecurity awareness.

TABLE 7.1
Proposed New Biosecurity Levels (BSec-Ls)

Management			
BSec-L1 Features	**BSec-L2 Features (in Addition to BSec-L1)**	**BSec-L3 Features (in Addition to BSec-L2)**	**BSec-L4 Features (in Addition to BSec-L3)**
• **Policy approval**: The organization has a general policy on biorisk management that has been developed, authorized, and signed by the organization's senior management • **Assessment**: Have a system to conduct and review biosecurity assessments • **Revisions**: Biosecurity policy is revised periodically, based on risk assessment and ongoing projects	• **Policy involvement**: Senior management is actively involved in drafting, executing, and managing the biosecurity policy • **Budget**: Have an adequate budget and other resources allocated for the management and implementation of a biosecurity program • **Surveillance**: Have a system to monitor unauthorized personnel that allows them to conduct routine nonlaboratory functions • **Assessment**: Bolster the system to more regularly conduct and review the biosecurity assessments • **Supervision**: Assign personnel to oversee the implementation of biosecurity measures • **Frequency**: Conduct periodical risk assessments on dual-use and other technological advances	• **Stricter supervision**: Adherence to the procedures and rules of conduct are being strictly monitored • **Counterfeits**: Have a system to mitigate the risk for counterfeit items. For example, having a list of approved vendors that provide reliable equipment and is revised periodically • **Vendor selection**: Have an approved list of certified vendors and buyers for biological substances and other biosecurity related equipment	• **Same as BSec-L3**

(Continued)

TABLE 7.1 (*Continued*)
Proposed New Biosecurity Levels (BSec-Ls)

Biosecurity Awareness			
BSec-L1 Features	**BSec-L2 Features (in Addition to BSec-L1)**	**BSec-L3 Features (in Addition to BSec-L2)**	**BSec-L4 Features (in Addition to BSec-L3)**
• **Increase awareness**: Have annual biosecurity awareness activities for all personnel working in the laboratory and support staff (IT, Facilities, Security, etc.) • **Orientation**: Have an entry-level biosecurity orientation program for new personnel, explaining the priority areas in biosecurity, and their roles and responsibilities • **Qualified officers**: The Biosecurity program transcends through the organization, and is it run by knowledgeable personnel • **Response protocol**: Personnel are aware of the existence of a response mechanism and their roles and responsibilities in case of any biosecurity breach • **Individual responsibilities**: Personnel are informed of their responsibilities regarding biosecurity and how responsibilities are assigned	• **Ongoing program**: Have a continuous training program planned for all personnel implementing biosecurity, including response to biosecurity breaches • **Qualified officers**: The Biosecurity program is sustainable throughout the organization, and it is run by highly qualified personnel • **DURC training**: Dual-use (and other technology advances) awareness is being incorporated into the training and awareness programs • **Reporting a breach**: Personnel are aware of the reporting mechanism for any biosecurity breaches and that the whistleblower's anonymity is protected • **Response protocol**: Personnel are competent in their knowledge of the response mechanism and their roles and responsibilities in case of any biosecurity breaches	• **Response protocol**: Personnel are competent in their response SOPs and their roles and responsibilities in the event of any biosecurity breach. They train often. The frequency should be determined based on the size of the staff, their level of proficiency on the protocol, staff turn-over, and other unique factors as determined by a biosecurity risk assessment	• **Response protocol**: Personnel train regularly to maintain competency. The frequency should be determined based on the size of the staff, their level of proficiency on the protocol, staff turn-over, and other unique factors as determined by a biosecurity risk assessment

(*Continued*)

TABLE 7.1 (*Continued*)
Proposed New Biosecurity Levels (BSec-Ls)

Physical security			
BSec-L1 Features	**BSec-L2 Features (in Addition to BSec-L1)**	**BSec-L3 Features (in Addition to BSec-L2)**	**BSec-L4 Features (in Addition to BSec-L3)**
• **Security level**: There is a basic level of security, such as doors and windows that lock the space outside of regular working hours. • **Restrict access**: Management enforces an access control policy	• **Security level**: There are varying degrees of security in different areas of the facility, including doors and windows that are always locked, and surveillance cameras placed in strategic places of higher vulnerability • **Stricter access**: Management enforces a stricter access control policy • **Monitoring**: Access controls are monitored for each secured area • **Unauthorized access**: There is an intrusion detection system to detect unauthorized entry to the facility and biological agents' storage areas • **Self-closing doors**: The laboratory doors are self-closing • **Building access**: Locks and keys to all buildings and entrances are supervised and controlled by a control official. Keys and badge access are issued only to authorized personnel	• **Security level**: There is camera coverage for all exterior laboratory building entrances. Alternatively, there is a guard walking the perimeter of the lab area • **Identification**: Badges are used to identify all personnel and visitors within the confines of the controlled areas. Depending on resources and geographic location, security guards at the laboratory entrances are an option instead of badges • **Escorting visitors**: There is a visitor escort procedure established for designated secured areas • **Building access**: Locks and keys to all buildings and entrances are supervised and controlled by a control official. Keys and badge access are issued only to authorized personnel after passing strict background checks	• **Escorting visitors**: Visitors are not allowed in these areas, with very few exceptions for building/ equipment maintenance, inspectors, and with approval from the facility director. There is a visitor escort procedure established for such cases

(*Continued*)

TABLE 7.1 (*Continued*)
Proposed New Biosecurity Levels (BSec-Ls)

BSec-L1 Features	BSec-L2 Features (in Addition to BSec-L1)	BSec-L3 Features (in Addition to BSec-L2)	BSec-L4 Features (in Addition to BSec-L3)
Physical security			
		• **VBM storage**: Valuable biological materials (VBM) are stored in secured locations • **Material inventory**: VBM are inventoried and regularly reconciled and audited • **Stricter access**: The entrance to the secured area or storage location is secured by combining various methods, including multiple keys and badge access	• **Stricter access**: Storage location is secured by combining various methods, including multiple keys and badge access. Three factor verification: (1) something you have, (2) something you know, and (3) something you are. The configuration is such that it requires a buddy system: at least two people together to be able to enter the spaces and access where VBM or select agents are either stored or used (both should be secured)
Materials accountability			
• **VBM policy**: Have a policy on the inventory management of valuable biological materials. • **Investigation protocols**: In the event of unusual or suspicious events, the organization has a system that triggers an investigation	• **Transfer policy**: Have a policy on the transfer of valuable biological materials • **Check in**: There is biosecurity signage at the entrance of laboratories and storage spaces that indicate the presence of biological agents without revealing the organisms	• **Preventive measures**: Have biosecurity procedures to prevent the deliberate dispersion of biological agents • **VBM management**: There is a designated person responsible for the registration and the active management of VBM to safeguard the control of these materials	• **Same as BSec-L3**

(*Continued*)

TABLE 7.1 (*Continued*)
Proposed New Biosecurity Levels (BSec-Ls)

Materials accountability			
BSec-L1 Features	**BSec-L2 Features (in Addition to BSec-L1)**	**BSec-L3 Features (in Addition to BSec-L2)**	**BSec-L4 Features (in Addition to BSec-L3)**
	• **Up-to-date inventory**: Maintain and update inventory records regularly	• **Up-to-date inventory**: Maintain and update inventory records regularly. The organization conducts periodic reviews of biological agents' inventory. Inventory storage locations are minimized and adequate. Protection is provided so that only authorized personnel have access • **Detailed inventory**: The inventory system in the organization includes detailed information regarding the location of the biological agents • **VBM amounts**: Valuable biological materials at the organization are limited to a certain quantity set by either international/ national or institutional policy	

(*Continued*)

TABLE 7.1 (*Continued*)
Proposed New Biosecurity Levels (BSec-Ls)

Information security			
BSec-L1 Features	**BSec-L2 Features (in Addition to BSec-L1)**	**BSec-L3 Features (in Addition to BSec-L2)**	**BSec-L4 Features (in Addition to BSec-L3)**
• **Specific policy**: There is a developed and implemented policy on information security. They are enforced regarding individual authorizations and their access to sensitive or confidential information • **Passwords**: Computers that store sensitive or confidential information are always password protected • **Network**: All computers are on the institutional network • **Unauthorized software**: Personnel cannot download and install unauthorized programs as they may contain malware • **Phishing scams**: Routinely evaluate the employee's vulnerability to phishing scams	• **Classification**: Use a classification system to rank the level of sensitive information • **Sensitive information access**: All personnel know and understand the procedures for accessing sensitive or confidential information • **Responsibility roles**: Have assigned authorized personnel who are responsible for information security • **Admin-level access**: Individuals have limited administrative access to their computers • **Catching vulnerabilities**: Run routine exercises to assess vulnerabilities • **Security software**: Install relevant security software on computers that store sensitive or confidential information • **Backups**: Have an information backup system for sensitive or confidential information • **Breach response**: Have emergency response procedures in the event of a breach of information security	• **Safe storage**: Sensitive or confidential information, including paper information, is stored in a physically secure place • **Restrict access**: Individuals do not have administrative access to their computers	• **Same as BSec-L3**

(*Continued*)

TABLE 7.1 (*Continued*)
Proposed New Biosecurity Levels (BSec-Ls)

BSec-L1 Features	BSec-L2 Features (in Addition to BSec-L1)	BSec-L3 Features (in Addition to BSec-L2)	BSec-L4 Features (in Addition to BSec-L3)
	• **Sharing sensitive information**: Implement administrative control measures for the exchange of sensitive information within and between different organizations		
Transport security			
• **Transport protocols**: There are personnel responsible for transporting valuable biological materials who are trained in specific requirements and procedures for transporting these materials	• **Selected carriers**: Have a preselection procedure for the transport companies they intend to use for the transportation of valuable biological materials. • **Compliance**: The selected transport company must be vetted and comply with local, national, and international legislation. • **Agreement**: There is a material transfer agreement between the organization and the sender/recipient of the valuable biological materials. • **Assessing transport risks**: Conduct risk assessments for each transportation type used.	• **Designed custody**: Enforce chain of custody. • **Track and trace**: There is a system to track and trace the transportation of biological samples available. • **Biosecurity at the receiving organization**: The organization ensures that the recipient institution has the appropriate level of biorisk management to receive and work with the sample before sending the sample. • **"Lost and theft" policy**: There is an emergency response plan to address the risk of packages being lost during transportation.	• **Same as BSec-L3**

(*Continued*)

TABLE 7.1 (*Continued*)
Proposed New Biosecurity Levels (BSec-Ls)

Personnel Reliability			
BSec-L1 Features	**BSec-L2 Features (in Addition to BSec-L1)**	**BSec-L3 Features (in Addition to BSec-L2)**	**BSec-L4 Features (in Addition to BSec-L3)**
• **Personnel readiness**: Have a personnel assessment system • **Onboarding protocols**: New people in the organization are subjected to a formal background screening process, including credentials, skills, personnel traits, and relevant background checks based on risk assessment. • **Reporting protocols**: Have policies and guidelines for personnel to report or register unusual behavior in coworkers or visitors. • **External monitoring**: Have SOPs or guidelines to monitor employees working outside regular hours	• **Frequent background checks**: Background checks are conducted on all personnel at least every 10 years or on a case-by-case basis after "suspicious activities." What the background checks consist of should be determined by institutions in collaboration with local/state/national governments. • At a minimum, I propose that they should include verification of educational qualifications, criminal background checks, and fingerprints entered into a database for law enforcement surveillance • Note: Background checks should not be used punitively after an incident nor in retaliation for whistleblowing • **Mental health assessment**: Mental health assessments or psychological assessments are conducted prior to employment	• **Afterhours work**: Restrict work outside regular hours • **Frequent background checks**: Background checks are conducted on all personnel at least every 5 years. The background checks should expand to include mental health/well-being checks • **Mental health assessment**: Mental health assessments or psychological assessments are conducted periodically during employment • **Visitor policies**: Visiting personnel (students, contractors, visitors, clients, temporary workers, etc.) must undergo a security clearance to access the facility • **Removing/reducing access**: Personnel transferred out of areas of increased risk must have their access controls immediately revoked	• **Afterhours work**: Forbid work outside regular hours except during emergencies and only if supervised. Implement a buddy system as needed • **Mental health assessment**: Mental health assessments or psychological assessments are conducted every 5 years during employment • **Visitor policies**: Visiting personnel (students, contractors, visitors, clients, temporary workers, etc.) are not allowed access to the facility, with rare exceptions, such as equipment maintenance and facilities repairs

(*Continued*)

TABLE 7.1 (*Continued*)
Proposed New Biosecurity Levels (BSec-Ls)

Personnel Reliability			
BSec-L1 Features	**BSec-L2 Features (in Addition to BSec-L1)**	**BSec-L3 Features (in Addition to BSec-L2)**	**BSec-L4 Features (in Addition to BSec-L3)**
	• **Visitor policies**: Have a policy and guideline for visiting personnel (students, contractors, visitors, clients, temporary workers, etc.) regarding security clearance to access the facility • **Authorized access list**: Maintain an up-to-date list of personnel with authorized access to the facility and biological agents • **Personnel relocation**: Have a system to assess when existing personnel are transferred to areas with an increased risk profile • **Removing/reducing access**: Have a system for the removal and exclusion of personnel (both temporary and, if appropriate, permanent) from access to the facility or access to the biological agents where it deems necessary through risk assessment		

(*Continued*)

TABLE 7.1 (*Continued*)
Proposed New Biosecurity Levels (BSec-Ls)

Emergency Response			
BSec-L1 Features	**BSec-L2 Features (in Addition to BSec-L1)**	**BSec-L3 Features (in Addition to BSec-L2)**	**BSec-L4 Features (in Addition to BSec-L3)**
• **Response plan**: Have an emergency response plan to respond and control biological emergencies or a biosecurity breach effectively • **Response plan content**: The emergency response plan contains tasks, responsibilities, and authorizations for response and recovery, including investigating biological incidents or emergencies	• **Contingency plan**: Have a contingency plan to guarantee the continuation of day-to-day operations with a sufficiently high level of security • **Third parties**: Have protocols involving relevant third parties (such as local health departments and local FBI office) in the event of a biosecurity breach or emergency • **Drills**: Conduct emergency drills or exercises that include biosecurity risks to determine that personnel can respond adequately to emergencies and other biosecurity situations according to plans or expectations • Establish procedures to correct situations where biosecurity is compromised • Establish preventive actions or revise procedures to ensure that breaches or biosecurity emergencies will not recur	• **Same as BSec-L2**	• **Same as BSec-L3**

- **Biosecurity Level 2 (BSec-L2)**: This level is for work with synthetic biology agents or techniques that pose moderate-to-high biosecurity risks. It builds on Level 1 with additional measures like restricting access to only authorized persons, enhanced inventory controls, and increased physical security protections. Work at this level requires biosecurity risk assessments and the implementation of commensurate biosecurity procedures.

- **Biosecurity Level 3 (BSec-L3)**: This level is for work with synthetic biology agents or techniques that pose serious or potentially lethal or severely damaging biosecurity risks. It requires stringent access controls (potential security clearance and background checks), rigorous inventory controls (including regular inventory reconciliation and audits), comprehensive personnel reliability checks, advanced physical and cyberbiosecurity protections (such as stringent controls over data access, camera surveillance, etc.), and detailed biosecurity plans. Work at this level warrants close oversight and regulation to mitigate biosecurity risks.
- **Biosecurity Level 4 (BSec-L4)**: This level is reserved for work with the highest risk synthetic biology agents and techniques that pose catastrophic biosecurity risks. It requires maximum biocontainment precautions, restricted access to only essential project personnel, stringent security screening, advanced access controls and surveillance, and extensive biosecurity audits. Work at this level necessitates the highest level of oversight and regulatory control to prevent misuse.

The key aim across all levels is to implement a graded series of precautionary measures calibrated to the specific biosecurity risks associated with synthetic biology research, thus balancing scientific freedom, performing responsible actions, verified by careful oversight by competent authorities. The proposed levels should include Select Agents (SA) regulations for relevant laboratories; however, I argue these should be applied to laboratories working with synthetic biology regardless of whether SA rules apply. These BSec levels allow standardizing biosecurity practices appropriate for different types of work.

The proposed Biosecurity Levels would provide a structured, scalable approach to biosecurity risk management, helping to ensure that appropriate measures to prevent, detect, and respond to potential biosecurity threats in synthetic biology laboratories. Implementing BSec-Ls would encourage a culture of security awareness, accountability, and responsibility, thus fostering safer and more secure research environments.

7.11 BEST PRACTICES FOR LABORATORIES & RESEARCH INSTITUTIONS

Implementing best practices in laboratories and research institutions is critical for establishing and maintaining biorisk management, ensuring regulatory compliance, and fostering responsible conduct of research. The IBC is key in defining, implementing, and overseeing these practices. Some best practices for laboratories and research institutions include:

- **Risk Assessment**: Every research project involving potentially biohazardous material should undergo a thorough risk assessment. This should identify potential hazards, evaluate the likelihood and potential impact of exposure, and define appropriate risk management measures. Figure 7.1 is a graphical representation of a matrix used to determine containment levels based on risk. The graph is color-coded for clarity:
  - **Light Gray**: The light gray area outlines the core requirements inherent to any laboratory environment. These requirements include hand washing sinks, doors, closable windows, and in certain risk scenarios, HVAC systems with negative or neutral air pressure compared to surrounding spaces. Additionally, a baseline level of PPE (i.e., lab coats, eye protection, and gloves) is mandated, along with a waste management system.
  - **Medium Gray**: The medium gray zone signifies heightened control measures, which are contingent upon a risk assessment and are not standardized. Examples of such measures might include enhanced PPE like N95 or PAPR respirators, stricter protocols, or specific facility modifications like negative air pressure environments.

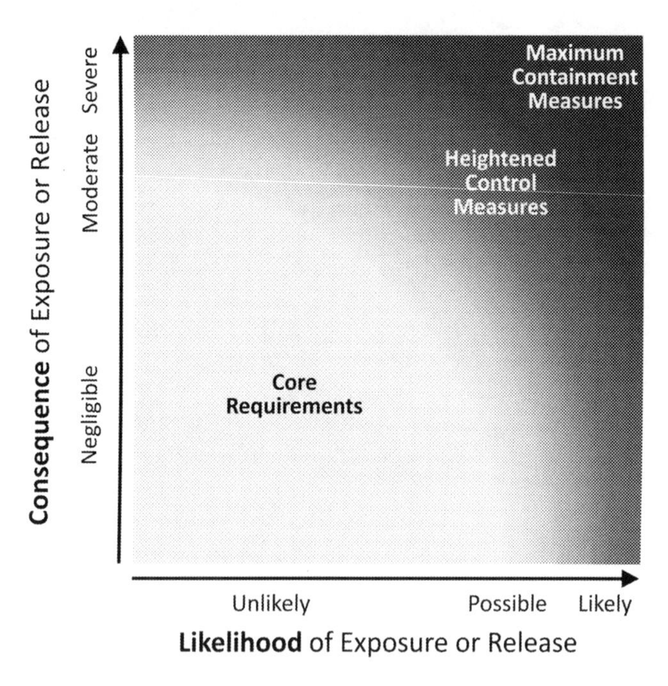

FIGURE 7.1 A risk assessment matrix depicted in graphical form to help assess containment levels in a laboratory setting. Containment levels are determined by assessing the likelihood and consequences of exposure or release of valuable biological material.[67] The likelihood of such an event can range from being unlikely to possible to likely, while the consequences can be categorized as negligible, moderate, or severe. A graphical representation depicts the containment levels, starting from core requirements suitable for low to medium risk levels (light gray), transitioning to heightened control measures for specific risk levels or pathogens (medium gray), and culminating in maximum containment levels for the most severe risks or pathogens (dark gray).

- **Dark Gray**: The dark gray zone denotes the strictest containment level, implemented when the potential exposure or release is deemed highly likely, and the consequences are profoundly severe. This could be pertinent for novel pathogens with high mortality and morbidity rates and no known treatment or cure. Such a scenario requires extreme biosecurity measures, and employees working in these high-risk areas often undergo thorough background checks or another form of personnel reliability check.

To effectively use this chart, begin by plotting the likelihood against the consequence of the intended work, providing a visual aid to assess the requisite containment level. A supplementary tool, the BiosecurityRAM software,[155] offers semiquantitative values for this chart. Depending on where your activity falls on the chart, appropriate containment measures are determined. For tasks falling under the maximum containment measures, a comprehensive risk-benefit analysis is vital to decide if the research is worth pursuing. If it is determined that the benefits outweigh the risks, a rigorous and lengthy permitting process ensues, encompassing local, state, and federal regulatory permissions. This is accompanied by extensive engineering controls, facility modifications, heightened PPE, and stringent biosecurity protocols.

- **Biosecurity Levels**: Every research project should conduct a risk assessment and, based on this determination, then derive the biosecurity level for the proposed work. Combined with a biosafety risk assessment, the Biosecurity Level should define the lab operating environment (Figure 7.2).
- **Training**: All personnel should receive biorisk management training appropriate to their roles and the risk associated with their work. This training should be refreshed regularly and whenever new risks are introduced.
- **Waste Management**: Biohazardous waste should be properly managed to prevent accidental exposure or release. This usually involves inactivating the material (e.g., by autoclaving) before disposal.
- **Physical Security Management**: The physical security of a laboratory working with synthetic biology is an integral part of its biosecurity and biorisk management plan. Physical security measures are intended to prevent unauthorized access to the lab and its contents, including biological materials, data, and equipment.

  A well-structured Physical Security Management plan should encompass a combination of physical barriers and procedural controls. These measures should be proportionate to the potential risks identified in the laboratory's risk assessment.

  Physical barriers can include lab design features such as fences, security guards, secured doors, locks, walls, and windows. Labs working with high-risk agents or techniques may also consider additional security measures such as controls, security cameras, and alarmed entrances.

  Inside the lab, biological materials and sensitive data should be securely stored. This should include locked storage cabinets for biological samples and secure servers or encrypted drives for data. Workstations should be

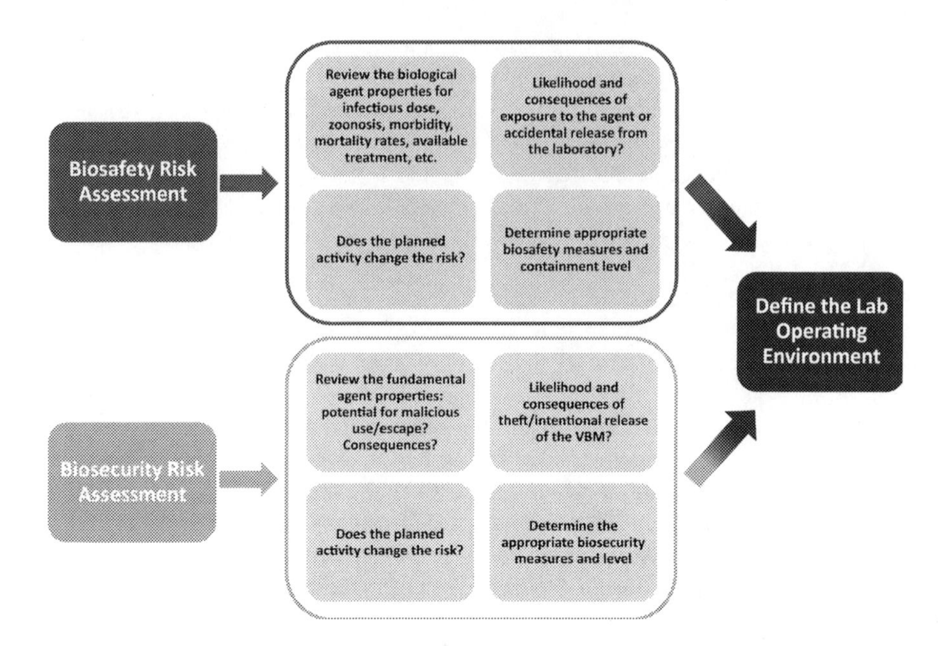

FIGURE 7.2 Biosafety and biosecurity risk assessments combined define the lab operating environment. An effective biorisk management system requires a comprehensive assessment of biosafety and biosecurity risks within the lab setting. The biosafety risk assessment evaluates the properties of biological agents, the probability and outcomes of unintended exposure or release, and the intended activities. This analysis helps determine the risk within acceptable limits and identifies appropriate measures to minimize it. Conversely, a biosecurity risk assessment focuses on the properties of the agent or Valuable Biological Material (VBM), the potential for its intentional misuse, and the repercussions of such actions. The likelihood and consequences are weighed against the planned activities to decide on suitable biosecurity precautions. This flowchart underscores the importance of a systematic approach to risk evaluation in laboratory settings.

clear of sensitive information when not in use, and electronic devices should require a password or biometric authentication.

Procedural controls form another layer of physical security. These include policies and procedures for lab access, visitor management, key control, and emergency response. Regular training for lab personnel on these policies and procedures is essential to maintain a high level of security awareness and compliance.

Physical Security Management is not a static process; it should be regularly reviewed and updated to reflect changes in the lab's activities, personnel, or the broader threat environment. It is also crucial to ensure that physical security measures do not impede the lab's functionality or the safety of personnel, thus striking a balance between security, safety, and operational efficiency.

- **Data Security Management**: In synthetic biology, data security is as crucial as physical security, if not more so. Information and data generated from synthetic biology research can be highly sensitive and valuable, whether it is the sequence of a novel organism, the structure of a protein, the dataset used to train an AI or the methodology of a new technique. If such information falls into the wrong hands, it can have severe implications for biosecurity, potentially enabling misuse or biological attacks. Therefore, maintaining data security is paramount. Effective Data Security Management in synthetic biology should include the following key components:

- **Access Control**: Limiting who has access to sensitive data is one of the first steps in securing data. Not everyone in an organization needs access to all data. Implementing role-based access control can restrict access to sensitive data only to those who need it to perform their job.
- **Data Encryption**: Encrypting data, especially when transmitted or stored, can protect it from unauthorized access. Even if data is intercepted or a storage device is lost or stolen, encrypted data will remain inaccessible without the correct decryption keys.
- **Secure Data Storage**: Data should be stored securely, whether on-site or in the cloud. This includes physical security measures for on-site servers, as well as virtual security measures for cloud storage, such as secure and private cloud environments.
- **Data Backup and Recovery**: Regular data backups can help recover lost data in a data loss incident. Moreover, having backup data stored in a separate, secure location can prevent total data loss in case of physical damage to the primary storage site.
- **Network Security**: Networks, both wired and wireless, should be secured to prevent unauthorized access. This includes using firewalls, secure network protocols, and regularly updating and patching network hardware and software to protect against known vulnerabilities.
- **Monitoring and Audit**: Regular monitoring and auditing of data access and usage can help detect and respond to suspicious activities or data breaches. Automated monitoring tools can help flag unusual behavior that might indicate a security incident.
- **Data Sharing Protocols**: When sharing data with external collaborators or publishing data, appropriate measures should be in place to protect sensitive information. This can include anonymizing data, using secure data transfer methods, or using secure data-sharing platforms.
- **Training and Awareness**: Users are often the weakest link in data security. Regular training and awareness programs can ensure that all personnel understand the importance of data security and follow the best practices to maintain it.

The dynamic nature of cyber threats means that data security management is an ongoing process that needs to keep adapting to new challenges. It is essential to regularly review and update data security measures considering new threats and vulnerabilities, technological advances, and changes in the organization's operations or regulatory environment.

- **Emergency Response**: Laboratories should have procedures to respond to accidents or emergencies, such as theft, breach of security, intentional exposures, or intentional equipment failures.

- **Personnel Management**: The human element is a vital component of a laboratory's biosecurity framework. People are often seen as the weakest link in security chains, but with proper management and training, they can become the strongest asset in maintaining robust biosecurity.

  In synthetic biology, Personnel Management should be multifaceted, incorporating elements such as background checks, security clearances, and continuous monitoring for personnel with access to sensitive materials or data. This is to ensure that the individuals handling high-risk biological materials are trustworthy and have the necessary knowledge and skills to do so safely and securely.

  Background checks can help assess an individual's trustworthiness before they are granted access to the laboratory and its resources. These checks can vary in scope and depth, depending on local regulations and the potential risks associated with the work. They may include criminal record checks, verification of educational credentials, and reference checks.

  Security clearances are used to further control access to specific sensitive or high-risk areas, materials, or data. The process for obtaining a security clearance often involves a more extensive background check, and clearances should be periodically reviewed and updated, as necessary.

  Regular monitoring of personnel is another critical aspect of Personnel Management. This includes not only surveillance to prevent and detect unauthorized activities, but also ongoing assessments of employees' competency and adherence to safety and security protocols. It also includes maintaining a culture of security awareness and responsibility, where employees feel empowered to report any potential security concerns or breaches.

  Personnel Management should also consider the "insider threat" – the risk that an employee or contractor could intentionally or unintentionally cause harm. This can be mitigated through a combination of comprehensive training, employee assistance programs, and a supportive workplace culture that encourages adherence to ethical standards and reporting of concerning behaviors.

  These measures are not meant to create a culture of suspicion, but rather one of trust, respect, and shared responsibility for biosecurity. The goal is to ensure that everyone in the laboratory understands and is committed to their role in maintaining biosecurity.
- **Inventory Management**: Proper management and surveillance of the inventory of biological materials, reagents, and associated data is crucial in synthetic biology research. A robust inventory management system not only ensures smooth laboratory operations but also significantly contributes to enhancing biosecurity.
  - **Inventory**: It is an integral part of a lab's biosecurity plan. It helps track the usage of sensitive biological materials and ensures they are handled and stored appropriately to prevent any misuse, accidental release, or biosecurity breach. Regularly updating and reviewing the inventory can help detect unauthorized access or unusual activity. Inventory logs can also provide valuable forensic evidence in the case of a biosecurity incident.

- **Inventory Control**: A systematic approach to inventory control can help maintain optimal levels of various items, prevent overstocking or shortages, and enable efficient use of resources. Advanced technologies such as barcoding or Radio Frequency Identification (RFID) can facilitate accurate and efficient tracking of inventory items. Inventory control measures also play a critical role in minimizing the chances of misplacing biohazardous materials or losing track of them.
- **Management**: Effective inventory management involves assigning clear roles and responsibilities for inventory control, standardizing procedures for receiving, storing, and discarding materials, and ensuring compliance. It also includes maintaining updated records of the source, current location, and disposal status of every item. Inventory management software can automate many of these tasks, reducing errors and increasing efficiency.
- **Inventory Reconciliation**: Inventory reconciliation is a critical part of biosecurity management in laboratories dealing with synthetic biology. With numerous biological agents, equipment, and other resources utilized in these settings, keeping track of these items becomes essential for maintaining order, efficiency, and security.

Inventory reconciliation involves a regular, thorough check of the actual inventory against the recorded inventory. This practice helps ensure that all biological agents and materials are accounted for and that discrepancies are swiftly identified and addressed. For example, if an organism is missing or extra quantities of material are found, an immediate investigation can be launched to find out why.

This process can be facilitated using inventory management software that allows real-time tracking of all items. It helps to have barcoded or RFID-tagged items to make it easier to log and track them. In synthetic biology, specific considerations might include:

- **Live Organisms**: Genetically modified organisms should be meticulously tracked. This includes noting the strain, the modifications made, and the storage conditions required.
- **DNA Sequences**: Synthetic DNA sequences should be carefully cataloged and stored, especially since they may pose a biosecurity risk if they were to be misused.
- **Equipment**: Tools and machines, particularly those used in the creation or modification of organisms, should also be included in the inventory reconciliation process.
- **Waste Materials**: Waste materials, especially those containing genetically modified organisms or synthetic DNA, should be tracked until they are properly disposed of, to ensure they do not accidentally get released into the environment.

Conducting inventory reconciliation regularly not only maintains the accuracy of the inventory records, but it also allows for early detection of potential security breaches or procedural failures. It also aids in the optimization of resources, prevention of accidental loss, and compliance with local, national, and international regulations. A robust inventory reconciliation process, therefore, plays an essential role in promoting safety and biosecurity in synthetic biology laboratories.

  - **Surveillance**: Regular audits and inventory inspections help verify the accuracy of inventory records, identify any discrepancies, and assess compliance with inventory control procedures. Surveillance measures can also include monitoring storage conditions to ensure they meet the requirements for each type of biological material and taking corrective action if any anomalies are detected.
  - **Disposal**: A significant part of inventory management is the safe and secure disposal of biological materials after their usage. It is important to have a well-documented process for the disposal of biohazardous materials, ensuring that they do not pose a risk to the environment or biosecurity.

  Effective inventory management requires ongoing training of personnel, periodic review and updating of inventory control procedures, and adoption of new technologies and best practices. In synthetic biology labs, where new materials and data are continually being generated, robust inventory management is vital for maintaining biosecurity and enhancing operational efficiency.

- **Auditing and Review**: Regular audits should be conducted to ensure compliance with biorisk management practices. The results of these audits should be reviewed by the IBC and used to identify areas for improvement.
- **Metrics and Trends**: Keeping accurate records of all activities involving potentially biohazardous material is a critical component of any effective biorisk management program. These records, which may encompass everything from risk assessments, staff training, incident reports to waste disposal, form the basis for insightful metrics and trend analyses. These analyses help to evaluate the efficiency of the system and its components, assess its effectiveness, and identify areas for improvement.
  - **Predicting System Weaknesses**: By regularly collecting and analyzing data, potential weaknesses in the system can be identified before they lead to incidents. For instance, if records show that specific types of incidents are occurring more frequently or that certain protocols are often being violated, these may be signs of underlying weaknesses in the system, such as inadequate training, unclear procedures, or lack of necessary resources.
  - **Evaluating Mitigation Measures**: The effectiveness of mitigation measures can also be evaluated through metrics and trend analyses. For example, if introducing a new safety protocol or training program leads to a decrease in incidents, this would be evidence of its effectiveness. Conversely, if no improvement or even a worsening trend is observed after implementing a certain measure, it may need to be revised or replaced.
  - **Continuous Improvement**: Metrics and trend analyses allow for continuous improvement of the biorisk management program. By identifying trends and patterns in the data, laboratories can anticipate potential issues, adapt their practices accordingly, and measure the impact of their interventions. This continuous feedback loop fosters a culture of continuous learning and improvement, which is crucial for managing the evolving risks in synthetic biology.

- **Benchmarking**: Besides monitoring internal trends, laboratories should also use metrics for benchmarking their performance against other institutions or industry standards. This can provide further insights into best practices and areas for improvement.

The collected data and resultant metrics are a powerful tool for managing biosecurity risks in synthetic biology. However, it is important to ensure that the data collection process itself adheres to privacy regulations and that the data is securely stored to prevent unauthorized access.

By implementing these best practices, laboratories and research institutions can mitigate the risks associated with synthetic biology, protect lab personnel and the public, and contribute to the responsible conduct of research.

7.12 CONCLUDING THOUGHTS

The emphasis on biorisk management in synthetic biology is necessary given the field's rapid evolution and the potential risks associated with its advancements. Comprehensive biorisk management practices are indispensable to navigating the complex dynamics between groundbreaking scientific discoveries and their potential risks. A strong focus on "Safe by Design" principles, which involve integrating safety considerations into every aspect of synthetic biology research, is underscored. The balance between ensuring safety and security while advancing scientific innovation requires careful attention. There is a critical need for robust biorisk management systems, especially considering the fast-paced development of synthetic biology and its potential implications for health, safety, and the environment. I recommend adopting a holistic approach. Continuous adaptation and responsiveness of biorisk management strategies are essential to meet synthetic biology's ever-evolving challenges and opportunities without stifling its progress, thus ensuring that the field progresses responsibly and safely. The scientific community and policymakers must collaborate effectively to establish protocols that ensure biosecurity and ethical integrity. Such collaboration should also foster public trust and international cooperation, recognizing the dynamic nature of this field.

7.13 KEY TAKEAWAYS

- **Comprehensive Biorisk Management**: Stressing the importance of managing biological risks in synthetic biology.
- **"Safe by Design" Principles**: Integrating safety and security considerations into the design phase of synthetic biology projects.
- **Physical and Procedural Protocols**: Emphasizing containment and safety procedures in laboratory settings.
- **Ethical Considerations and Public Engagement**: Highlighting the need for ethical guidelines and engaging the public in discussions about synthetic biology.
- **International Cooperation**: Calling for global collaboration in establishing and following biorisk management standards.
- **Continuous Adaptation**: Recognizing the need for biorisk management strategies to evolve with advancements in synthetic biology.

7.14 THOUGHT-PROVOKING QUESTIONS

7.1 How should future biorisk management guidelines be written to ensure enough flexibility to accommodate updating the practices in response to future advancements in synthetic biology and possible new risks?

7.2 What are the factors that tend to be overlooked in the current practices of biorisk management in synthetic biology, both at the written policy level and during implementation in the laboratory setting?

7.3 What regulatory bodies should be responsible for drafting a new containment and disposal strategy when a new organism with entirely new synthetic genome is researched and developed? The research lab? The institution? Others? Should the research be allowed to start without an updated containment and disposal strategy? How does this potential bureaucracy slow down advancement in the field and government funding awards?

7.4 How frequently should regulations for synthetic biology be revised and updated to respond to emerging risks associated with innovative synthetic biology technologies?

8 Developing a Synthetic Biology Laboratory Biosecurity Manual

Up to this point in this book, the case has been made that biosecurity risks should be considered in synthetic biology, with case studies and regulatory frameworks summarized. This chapter provides a comprehensive guide for creating a biosecurity manual specifically tailored for laboratories engaged in synthetic biology research. It recognizes this field's unique challenges and risks and emphasizes the importance of a well-structured biosecurity manual to ensure safe and responsible research practices. The chapter then delves into developing and implementing the manual globally, nationally, and institutionally, stressing the importance of customization to address individual laboratories' specific needs and risks. It provides a step-by-step approach for drafting, reviewing, and updating the manual, ensuring that it remains relevant and effective in the ever-evolving field of synthetic biology. This chapter is a valuable resource for laboratory managers, researchers, and biosafety professionals in synthetic biology.

DOI: 10.1201/9781003423171-8

8.1 THE LABORATORY BIOSECURITY MANUAL

A Synthetic Biology Laboratory Biosecurity Manual is pivotal in assuring safe operations within synthetic biology (SynBio) laboratories and mitigating associated biosecurity risks. Such a manual should encompass the eight priority areas of biosecurity: management, awareness, physical security, materials accountability, information security, transport security, personnel reliability, and emergency response.[156,157] This detailed guide would tailor recommendations based on varying biosecurity risk levels associated with research areas, safeguarding normal operations in SynBio labs.

In order to create a comprehensive and effective Biosecurity Manual, it is suggested to engage a wide range of stakeholders, including national and local think tanks, lab representatives, international contributors, and individual experts spanning disciplines, such as scientists, ethicists, and philosophers. By incorporating these diverse perspectives, the Biosecurity Manual would be more inclusive, applicable, and progressive, ensuring its relevance in the rapidly evolving field of synthetic biology.

The Biosecurity Manual should cover a broad array of topics. It should include "General Principles," providing an overview of biosecurity in SynBio, and "Biosecurity Guidelines," detailing specific recommendations for different types of work and levels of risk. Other sections could deal with "Laboratory Technology and Equipment," outlining the safe use and maintenance of essential lab tools, and "Personnel Biosecurity Training," providing information on education and training programs to enhance lab personnel's understanding of biosecurity.

In addition to these measures, a well-functioning feedback system is crucial for the continued improvement of the Biosecurity Manual. This system should facilitate the flow of information from the ground level (lab workers, technicians, and researchers) up to the policy level. This feedback mechanism would capture insights from individuals directly engaged in lab work and facilitate necessary updates to the Biosecurity Manual based on their practical experiences.

Here, I propose an outline to start creating a Biosecurity Manual, as summarized in Table 8.1. I propose this manual based and improving upon previously published work that has been very helpful in this field.[158,159] However, I argue that we need to develop a more comprehensive and updated guidance document for global adoption. I encourage leaders and policymakers alike to engage with stakeholders and experts and develop this manual expediently, as SynBio is rapidly evolving, and scientists need expert guidance and guidelines.

Establishing clear guidelines for biosecurity in synthetic biology is critical. I propose developing and publishing a comprehensive Biosecurity Manual that integrates the measures discussed in this chapter to establish a balance that would maximize the development of synthetic biology while concurrently minimizing associated biosecurity risks. The Biosecurity Manual should serve as a valuable tool in navigating the intricacies of biosecurity in synthetic biology laboratories. It will provide a reliable, user-friendly framework that ensures safety and fosters innovation. This manual should be updated periodically by the authors. For a national or global manual, updates must be timely and coordinated with updates to biosafety and changes in related regulations. Similarly, institutional biosecurity manuals should undergo annual updates, in line with the practice for biosafety manuals.

TABLE 8.1
Outline of Key Elements in a Proposed Biosecurity Manual

Chapter Topic	Chapter Outline
Introduction	• Biosecurity and associated risks in a laboratory • Biosecurity and international obligations for laboratories • Biosecurity and national regulations for laboratories
Development of a laboratory biosecurity program	• Principles of biosecurity • Management of a laboratory biosecurity program • Biosecurity risk assessment • Laboratory biosecurity level criteria • Specific biosecurity recommendations
Principles of biosecurity	• Management • Biosecurity awareness • Physical security • Materials accountability • Information security • Transport security • Personnel reliability • Emergency preparedness and response
Management of a laboratory biosecurity program	• Roles and responsibilities • Biosecurity system design • Response force • Performance testing • Documentation • Assessments and audits • Training and exercises
Biosecurity risk assessment	• Overview biosecurity risk assessment methodology • Characterize assets and threats • Evaluate scenarios • Characterize the risk • Risk reduction
Laboratory biosecurity level criteria	• BSec-Ls 1–4
Specific biosecurity recommendation	• BSec-L1: Low-risk facility • BSec-L2: Moderate-risk facility • BSec-L3: High-risk facility • BSec-L4: Extreme risk facility
Appendix A	• Vulnerability and threat assessment questionnaires
Appendix B	• Biosecurity risk assessment methodology
Appendix C	• Biosecurity plan template
Appendix D	• Sample memorandum of understanding with local law enforcement
Appendix E	• Template SOP for testing access control systems • Template SOP for testing cyber-biosecurity systems
Appendix F	• Developing a robust auditing program for biosecurity

8.2 THE ROLE OF INTERNATIONAL ORGANIZATIONS

International organizations play a vital role in biorisk management in synthetic biology by providing leadership, education, and resources. These organizations operate across borders and bring together many stakeholders, making them well-suited to address the global and interconnected nature of biorisk management. Let us look at the roles of some key organizations:

- **World Health Organization (WHO)**: WHO stands as a global beacon for biorisk management, providing essential guidance and policy recommendations. It serves as a central resource in educating a broad audience, from policymakers to health practitioners and the public, about biosecurity risks and best practices. Additionally, WHO is committed to enhancing global capacity through training programs and providing vital resources, including the WHO Laboratory Biosafety Manual, with a focused effort on supporting low- and middle-resource countries.
- **Centers for Disease Control and Prevention (CDC)**: The CDC is a forefront authority conducting research and overseeing surveillance on biosecurity concerns, such as emergent infectious diseases of high consequence and biosecurity threats, and orchestrates responses to such incidents within the United States. It plays a critical role in disseminating biosecurity knowledge to the public, health experts, and policymakers. Furthermore, the CDC equips stakeholders with comprehensive resources and training on biosafety, biosecurity, and emergency preparedness, underpinning its commitment to safeguarding public health.
- **European Biosafety Association (EBSA)**: EBSA plays a pivotal role in managing biorisks within Europe's synthetic biology sector. It advances biosafety and biosecurity as scientific fields, encourages professional exchange, and facilitates continuous education. EBSA's core activities include developing safety guidelines for biological research and promoting safe practices in synthetic biology labs. It also offers professional development opportunities and fosters standardization of biosafety practices across Europe.
- **World Organization for Animal Health (WOAH)**: WOAH acts as the international authority on animal health, crucially involved in biorisk management within synthetic biology. It operates across national boundaries to enhance animal health and welfare globally. WOAH's strategies for managing biorisks include setting global standards, fostering education, and collaborating with organizations like WHO and FAO to ensure safe synthetic biology practices.
- **Food and Agriculture Organization (FAO) of the United Nations**: As the UN's leading agency against hunger, FAO collaborates with global partners, including WHO and WOAH, to promote a unified approach to biorisk management in synthetic biology. FAO's initiatives are vital for safeguarding health and the environment, providing leadership, knowledge, and resources to support safe and responsible synthetic biology activities.

By providing leadership, education, and resources, these organizations help to build capacity for biorisk management in synthetic biology, support the development of effective policies and practices, and foster a culture of safety and security in the scientific community and beyond. Their efforts are crucial for addressing the global and interconnected challenges of biorisk management in this rapidly advancing field.

8.3 MANDATING A RISK ASSESSMENT IN FEDERALLY FUNDED RESEARCH

The advent of synthetic biology has not only opened novel avenues in research but also raised new questions regarding biorisk management. This section explores the possibility of mandating a risk assessment in every federally funded grant related to synthetic biology, outlining its potential advantages and disadvantages.

Advantages

- **Enhanced Biosecurity**: One of the key benefits of mandating risk assessments is the proactive mitigation of potential biosecurity risks. By understanding the potential hazards upfront, researchers can build safeguards into their project designs, reducing the likelihood of inadvertent or deliberate misuse of biological materials.
- **Improved Research Quality**: Risk assessment encourages researchers to consider broader potential outcomes, fostering a more thorough, holistic approach to study design. This may improve research quality by preventing unintended negative consequences, ensuring that the generated knowledge does not pose an unnecessary risk to society.
- **Greater Public Trust**: Public trust in science is more critical than ever, and transparent risk assessments can demonstrate accountability and commitment to safety. This may reassure the public that their tax dollars are being used responsibly and that researchers are committed to minimizing potential risks associated with their work.

Disadvantages

- **Increased Administrative Burden**: Implementing mandatory risk assessments for every federally funded grant will undoubtedly increase the workload for researchers and administrators. This could potentially slow down the pace of research and increase the cost of performing research. However, this could be mitigated by adding biosecurity risk assessment as part of the already existing IBC process.
- **Complex Risk Assessment Process**: The process of conducting risk assessments in synthetic biology can be complex due to the novelty and unpredictability of certain research areas. The ability to predict all potential risks may be limited, resulting in incomplete or inaccurate assessments. Leveraging the expertise of the Institutional Biorisk Officer in this process,

combined with proper enforceable guidelines and regulation, could help mitigate this risk.
- **Hindrance to Innovation**: Some researchers may be dissuaded from pursuing specific research areas due to the perceived risk or the burden of conducting a complex risk assessment. This could potentially stifle innovation in critical areas of synthetic biology.
- **Misuse of Risk Assessment**: There is a potential risk that these assessments could be misused to halt research based on exaggerated or misrepresented dangers influenced by public fear or misunderstanding of synthetic biology. This risk can be mitigated by engaging stakeholders early in this process for riskier proposals and by continually raising public awareness and education on this topic.

Although mandating risk assessments in federally funded synthetic biology research brings additional challenges, those can be mitigated. The considerable benefits of enhanced biosecurity and improved research quality supersede the challenges in the author's expert opinion. The proper funding bodies and federal agencies should seriously consider them. This will be key in shaping a responsible and practical approach to advancing synthetic biology research.

8.4 TRAINING IN BIOSECURITY PRINCIPLES

Training in biosecurity principles is crucial to biorisk management in synthetic biology. Educating researchers on the potential risks and the practices to mitigate them can foster a culture of safety and responsibility in the field. Such training might involve:

- **Understanding Biosecurity Risks**: Training should start with an overview of the potential biosecurity risks associated with synthetic biology. This includes physical risks, such as accidental exposure to hazardous materials, and more abstract risks, such as misusing research for harmful purposes.
- **Containment Levels and Practices**: Researchers should be trained on different biosecurity containment levels (BSec-Ls) and the biosecurity practices associated with each level. They should understand the containment measures, safety equipment, and procedures for handling biohazardous materials.
- **Risk Assessment**: Researchers should be trained in conducting a risk assessment for their research, identifying potential hazards, assessing the likelihood and potential impact of exposure, and defining appropriate risk management measures.
- **Incident Response**: Training should cover how to respond in an accident or emergency, such as theft or misuse.
- **Ethical and Dual-Use Considerations**: Researchers should be aware of the ethical implications of their work, including the potential dual-use nature of synthetic biology research. They should understand the principles of

responsible conduct of research and the guidelines for publishing potentially sensitive research.

- **Regulations and Compliance**: Training should cover the relevant regulations and guidelines for synthetic biology research, including permit requirements, reporting, and oversight. Researchers should understand their responsibilities for compliance and the potential consequences of non-compliance.
- **Continuing Education**: Given the rapidly evolving nature of synthetic biology and biosecurity, training should not be a one-off event. Continuing education programs should ensure that researchers stay updated on new developments, best practices, and regulations.

By providing comprehensive training in biosecurity principles, research institutions can equip researchers with the knowledge and skills they need to conduct their work safely and responsibly, contributing to the overall management of biorisks in synthetic biology. Biosecurity Central already provides training and resources[160]; however, more need to be developed and made accessible.

8.5 COLLABORATIVE EFFORTS TO INCREASE PUBLIC AWARENESS

Public awareness and understanding of synthetic biology and its associated biorisks are essential for informed public debate, policymaking, and individual decision-making. Collaborative efforts between various stakeholders can significantly increase public awareness about this field. A few ways such collaborations might work include the following:

- **Collaboration between Researchers and Media**: Scientists and researchers can work with media outlets to ensure accurate and accessible reporting on synthetic biology. This might involve offering expert comments, helping journalists understand complex issues, or writing articles or opinion pieces for a general audience.
- **Town hall Meetings**: Establishing and maintaining open lines of communication with the public is a crucial component of effective biorisk management in synthetic biology. Town hall meetings serve as a valuable tool for achieving this goal, particularly when research institutions plan to conduct new experiments that pose significant risks to the local community, whether real or perceived.

  In synthetic biology, these public forums can help explain the science behind proposed experiments, clarify the actual risks involved, and explain the measures that will be taken to mitigate these risks. They provide an opportunity for researchers to present their work in a manner that is accessible and understandable to the public. Such proactive outreach can dispel misconceptions, alleviate concerns, and build trust between the research institutions and the communities they serve.

  Moreover, town hall meetings enable the residents to voice their concerns, ask questions, and offer suggestions. This two-way dialogue allows the public to feel heard and engaged. It is also an excellent opportunity for researchers to gain insights into local values, cultural perspectives, and potential area-specific issues that may affect the implementation of the project.

In situations where significant risks have been identified, these meetings serve as an avenue to negotiate conditions under which the research can proceed. It may include certain safety measures, emergency response plans, or community monitoring initiatives. Public consent is vital in such situations as it ensures social license for the research and fosters community acceptance and cooperation.

It is important to note that the conduct of these meetings should adhere to principles of transparency, inclusivity, and respect for diversity. They should be planned carefully for the local context, ensuring accessibility, and using language that is easy to understand for non-scientists.

- **Public Engagement Events**: Institutions involved in synthetic biology research can collaborate to host public engagement events such as open days, science festivals, or public lectures. These can be opportunities to explain the work being done, discuss potential benefits and risks, and answer questions from the public.
- **Educational Collaborations**: Partnerships between research institutions and schools or educational organizations can help to integrate synthetic biology into the curriculum, promoting early understanding of the field. This might involve developing educational materials, hosting workshops, or setting up mentorship programs.
- **Collaboration with Policymakers**: Scientists can work with policymakers to ensure they understand synthetic biology well and can make informed decisions about regulation and oversight. This might involve briefing sessions, expert testimony, or collaborative workshops.
- **Partnerships with Citizen Science Initiatives**: Involving the public in synthetic biology research through citizen science initiatives can increase understanding of the field and foster a sense of ownership and accountability.
- **Global Cooperation**: As synthetic biology and its potential biorisks are global issues, international cooperation is critical to increasing public awareness. This might involve sharing best practices for public engagement, coordinating public messages, or jointly responding to public concerns.

By working together, these different groups can help ensure the public has a balanced and informed understanding of synthetic biology, its potential benefits, and its associated biorisks. This can promote more informed decision-making, greater public trust in the field, and a more substantial societal consensus on managing it.

8.6 TECHNOLOGICAL SOLUTIONS TO MANAGE BIORISKS

Technological solutions form an integral part of managing biorisks in synthetic biology. These can range from containment systems and PPE to advanced computational models for risk assessment and AI-driven monitoring systems. Some technological solutions that could play a role in biorisk management include:

- **Detection and Monitoring Technologies**: Technologies for detecting and monitoring biohazards can help identify potential risks and respond swiftly to incidents. This could include biosensors for detecting specific pathogens or other biohazards or AI-driven systems for monitoring laboratory environments and identifying anomalies.
- **Computational Modeling**: Advanced computational models can be used to assess risks, predict the behavior of synthetic organisms, or simulate the spread of biohazards in the event of an accidental release. These models can inform risk management strategies and emergency response planning.
- **Data Security Technologies**: With the increasing digitalization of synthetic biology research, technologies for data security, such as encryption and blockchain, are important for protecting sensitive information and preventing cyberbiosecurity threats.

By leveraging these and other technological solutions, it is possible to build multiple layers of protection against biorisks in synthetic biology, enhancing the safety and security of this rapidly advancing field.

8.7 MONITORING AND SURVEILLANCE TECHNOLOGIES IN PUBLIC HEALTH

Monitoring and surveillance technologies in public health can be instrumental in identifying, tracking, and managing potential biohazards associated with synthetic biology. These technologies can facilitate real-time detection and rapid response to biosecurity events, helping prevent or mitigate public health harm. This approach is already being implemented worldwide by public health agencies in the campaign to eradicate polio. Here is how these technologies can be applied:

- **Biosensors**: Among the most promising applications of synthetic biology are biosensors, which are designed to detect specific biohazards such as pathogens or toxins. These sophisticated tools function by reacting to environmental substances or conditions, providing a targeted, rapid response system for potential threats. For instance, in public health, portable biosensors could monitor locally for outbreaks that could pose biosecurity risks. If such a risk is detected, the data could be transmitted to relevant authorities, ensuring that prompt and effective mitigation measures are activated. These risks include the sudden emergence of infectious diseases, accidental leakage of genetically engineered organisms from a lab, or the intentional release of harmful biological agents. In each case, the swift detection capability of biosensors can significantly enhance the capacity of health officials to respond, reducing potential damage and aiding in containment and treatment.

  In addition to their function as early warning systems, biosensors can also be instrumental in continuously monitoring known threats. This can be crucial for assessing the effectiveness of biosecurity protocols and identifying emerging trends and patterns. Integrating biosensors in biosecurity represents a significant step forward in our ability to manage biorisks associated with synthetic biology, enabling a far more responsive and preemptive approach to potential threats.

- **Genomic Surveillance**: Using genomic sequencing can help to track the spread of pathogens, identify new variants, and monitor the potential release of synthetic organisms into the environment. For instance, during the COVID-19 pandemic, genomic surveillance has been crucial in tracking the spread of the virus and identifying new variants.
- **Digital Epidemiology**: Digital tools have become indispensable in tracking and managing public health threats in the modern, interconnected world. For instance, data analytics applied to health reports across global health agencies, hospitals, and other health-related institutions can provide early warnings of emerging diseases or outbreaks. Such data collection might involve analyzing reported symptom patterns, identifying unusual increases in specific diseases, or detecting geographic clusters of illnesses. The aggregated information is valuable in identifying potential global health threats and can assist in swift, targeted responses to contain them. This digital epidemiology approach allows us to leverage large-scale data to enhance the real-time surveillance of public health, ensuring prompt and efficient responses to potential crises.
- **AI and Machine Learning**: Artificial intelligence and machine learning algorithms can analyze vast data, identify patterns, and predict potential threats. These tools can enhance public health surveillance's efficiency and effectiveness and help anticipate and respond to biosecurity threats.
- **Remote Sensing**: Technologies such as drones or satellites can monitor large or hard-to-reach areas for potential biosecurity risks, such as outbreaks of disease in wildlife populations that could potentially spill over to humans.
- **Health Information Systems**: Digital health information systems, such as electronic health records, can be valuable tools for surveillance, allowing public health officials to monitor trends in disease incidence and identify potential biosecurity threats.

By harnessing these and other technologies, public health officials can enhance their capacity to detect, track, and respond to biosecurity threats associated with synthetic biology, helping to protect public health and safety.

8.8 CONCLUDING THOUGHTS

Developing a stand-alone, globally adopted biosecurity manual for synthetic biology laboratories is crucial in advancing biosecurity in this rapidly evolving field. Beyond a mere regulatory requirement, this manual will be an indispensable tool to fill a clear gap for biorisk management professionals. The manual should function as a comprehensive guideline for maintaining biosecurity in different laboratory settings and as a valuable educational resource for biosecurity professionals. The manual's scope should be broad, addressing various facets of biosecurity, including biosecurity program development and implementation, biosecurity risk assessment, and more. It should prepare laboratory personnel for effective biosecurity incident response. Rapid technological advancements and emerging challenges characterize the field of synthetic biology. Keeping a biosecurity manual up-to-date will ensure that it remains relevant and practical. This dynamic approach to biosecurity underscores the importance of a well-structured, frequently revised manual that evolves with the field it serves. By doing so, the manual will become a living document reflective of the latest scientific advancements and best practices in biosecurity, thus playing a pivotal role in ensuring responsible and safe research in synthetic biology.

8.9 KEY TAKEAWAYS

- **Necessity of a Biosecurity Manual**: There is critical need for a specific and comprehensive stand-alone biosecurity manual in synthetic biology laboratories.
- **Comprehensive Coverage**: Importance of covering diverse biosecurity aspects such as program development, implementation, biosecurity risk assessment, and incident response.
- **Customization and Updating**: The manual must be tailored to specific biosecurity needs and regularly updated.
- **Stakeholder Engagement**: Engage a wide range of stakeholders, including scientists, biosecurity professionals, and policymakers in the manual's development.
- **Practical Application**: The manual should play a main role in guiding daily laboratory biosecurity practices and enhancing biosecurity awareness among staff.
- **International and Institutional Implementation**: It is important to implement the manual at global, national, and institutional levels.

8.10 THOUGHT-PROVOKING QUESTIONS

8.1 What regulatory body should be responsible for developing a probabilistic risk assessment model (flexible and adaptable) that incorporates the likelihood and impact of various biorisks? How frequently should this model be revised and updated to account for the rapid advancements in synthetic biology?

8.2 Which recent advancements in synthetic biology technologies are not currently addressed by the latest biosafety and biosecurity guidelines? How can these potential risks be effectively managed in the absence of appropriate policies?

8.3 What challenges might arise in implementing comprehensive biorisk management strategies in synthetic biology laboratories, and how can these be addressed?

8.4 What comprehensive guidelines should be included in a biosecurity manual to ensure the balance between advancing synthetic biology and minimizing biosecurity risks?

8.5 Besides offering leadership, education, and resources, what other roles should international organizations play in biorisk management, and how can they enhance global biosecurity standards while fostering global cooperation and standardization of biosecurity practices across borders?

8.6 How might mandated risk assessments for federally funded synthetic biology research impact innovation and public trust? Would the potentially increasing administrative burdens outweigh the enhancement in biosecurity and research quality and boost public trust in these mandatory risk assessments?

8.7 What are the challenges and benefits of integrating biosecurity training into synthetic biology research programs? Would fostering a culture of safety and responsibility outweigh all the efforts to stay current with evolving biosecurity concerns and technologies?

8.8 How can certain collaborative efforts, such as partnerships with media, public engagement events, and educational initiatives, help foster informed debate and decision-making among the public and policymakers, increasing public awareness and understanding of synthetic biology and its biosecurity implications?

8.9 Besides biosensors, computational models, and AI-driven monitoring systems, what other technological solutions are emerging to manage biorisks in synthetic biology, and how effective are they?

9 Incorporating Human Factors in Biosecurity for Synthetic Biology

Human factors are critical in establishing and maintaining the biosecurity of synthetic biology laboratories and research environments. This chapter highlights the necessity of recognizing and integrating human behavior and organizational culture into biosecurity strategies to enhance biosecurity in this rapidly evolving field. The chapter defines human factors and their relevance to biosecurity in synthetic biology. It discusses strategies for fostering a biosecurity-conscious mindset among researchers and staff, including effective communication, training programs, and leadership engagement in promoting biosecurity. The chapter also delves into organizational aspects, such as policy development, supervision, and establishing a reporting system for biosecurity incidents. It calls for a comprehensive approach to biosecurity that incorporates human factors. It argues that understanding and integrating the human element is crucial for effectively managing biosecurity risks in synthetic biology, leading to safer research practices and advancements in the field.

DOI: 10.1201/9781003423171-9

For a comprehensive reading in human factors in biorisk management, refer to the book *Prepare and Protect: Safer Behaviors in Laboratories and Clinical Containment Settings*.[161]

9.1 COMMUNICATION BREAKDOWNS

Effective communication forms the backbone of any successful operation within a laboratory setting, especially in synthetic biology, where the stakes concerning biosecurity can be incredibly high. Breakdowns in communication, resulting in miscommunication or misunderstandings, can lead to errors and accidents, jeopardizing the integrity of scientific research and potentially creating biosecurity risks.

Communication breakdowns can occur at various levels: between researchers within a team, between different teams within an institution, and institutions and oversight bodies. At each level, the nature and impact of communication breakdowns can vary, and therefore, each requires individual consideration and tailored strategies to prevent mishaps.

Within research teams, miscommunication can lead to errors in the execution of experiments or the misunderstanding of protocols, which can inadvertently create biosecurity risks. Clear and precise communication of experimental procedures, safety protocols, and biosecurity guidelines are critical. Regular team meetings, robust documentation, and open channels for queries and clarification can help mitigate this risk.

Between different teams or departments within an institution, breakdowns in communication can result in a lack of coordination or misunderstanding of roles and responsibilities. This can lead to gaps in biosecurity measures, especially when oversight of certain functions is assumed to be the responsibility of another team. Regular interdepartmental meetings, clearly delineated roles and responsibilities, and institution-wide communication channels can help prevent such issues.

At the institutional level, a lack of effective communication with oversight bodies or regulatory authorities can result in non-compliance with biosecurity guidelines or regulations, through ignorance or misunderstanding. Institutions must maintain clear lines of communication with these bodies, ensuring that all updates to regulations are understood and implemented promptly. Regular compliance training and strong relationships with regulatory bodies can assist.

The goal should be to foster a culture of open, transparent communication where all parties understand their roles and responsibilities and are comfortable asking questions or seeking clarification. Regular training programs, clear documentation, and robust communication channels can go a long way toward preventing communication breakdowns and the resultant risks. It is important to remember that effective communication is a two-way process, requiring both clear conveyance of information and active listening or comprehension.

Effective strategies to counter communication breakdowns, such as comprehensive training programs, regular team briefings, clear and open communication channels, and a no-blame culture that encourages reporting and learning from mistakes, are crucial. By recognizing the potential pitfalls of communication breakdowns and addressing them proactively, biosecurity in synthetic biology can be significantly enhanced.

9.2 COMPLACENCY

Complacency, characterized by overfamiliarity with a task or a belief in the inherent safety of a situation, poses a significant biosecurity risk in synthetic biology. When individuals become complacent, they may underestimate or disregard the potential dangers of their work. This lack of vigilance can lead to lapses in biosecurity measures, increasing the risk of accidents or biosecurity breaches. Complacency can stem from repetitive tasks, longstanding routines, or an environment where risks are perceived as minimal or absent. For instance, a researcher working regularly with low-risk organisms may become complacent and, consequently, less stringent with biosecurity measures. This can become problematic if the researcher begins working with higher-risk organisms without adjusting their practices. Moreover, complacency can breed in environments where biosecurity incidents are rare or nonexistent. In such cases, the lack of incidents can foster a false sense of security, leading individuals to believe that risks are minimal or that existing biosecurity measures are more than sufficient. It is essential to cultivate a culture of constant vigilance within the laboratory and the broader institution. This involves regular biosecurity training that emphasizes the potential consequences of complacency and underscores the importance of always maintaining stringent biosecurity practices.

Additionally, regular biosecurity audits and drills can help break the monotony of routine and instill the habit of constant vigilance. Audits (an inspection of an organization's processes/procedures/plans/documentation) can identify potential areas of complacency, uncover gaps in the current plans, and give a reasonable timeline to mitigate gaps uncovered, while drills or exercises can simulate biosecurity incidents, helping individuals understand the potential consequences of lax practices. Another effective strategy to combat complacency is rotating tasks where feasible. By periodically changing the tasks researchers are responsible for, it is possible to reduce the monotony that often breeds complacency, keeping individuals alert and mindful of their practices. Additionally, an added benefit is having a fresh pair of eyes on the task. Cross matrixing the people, skills, and work areas prevents becoming siloed, groupthink, and helps share good procedures across the institution. Further, fostering an environment where near misses and biosecurity breaches are reported and analyzed rather than overlooked or underplayed can help maintain awareness of potential risks. Such an environment encourages ongoing vigilance, promoting an understanding that complacency can have significant consequences. While complacency can pose a severe threat to biosecurity, it can be managed effectively with appropriate measures. By fostering a culture of vigilance, encouraging regular training, performing audits and drills, rotating tasks, and promoting an open reporting environment, it is possible to counteract the dangers of complacency and maintain high biosecurity standards in synthetic biology.

9.3 UNSAFE ACTS

Unsafe acts represent one of the most direct and significant threats to biosecurity in synthetic biology. These actions, whether performed intentionally or recklessly, break from the established safety protocols, and increase the risk of accidents, potentially leading to biohazards.

Unsafe acts can range from minor infractions, such as failure to wash hands after handling biological samples, to major breaches such as disabling safety systems, unauthorized access to restricted areas, or purposeful misuse of biological agents. These behaviors may include negligence, disregard for established protocols, inadequate training, or deliberate intent to cause harm.

To mitigate the risks associated with unsafe acts, it is crucial to foster a robust safety culture within laboratories and institutions. This involves several key components:

- **Training and Education**: Regular and comprehensive training programs are vital in informing employees about the consequences of unsafe acts, emphasizing adherence to safety procedures, and teaching the correct way to respond to potential biosecurity incidents.
- **Accountability and Enforcement**: Clear policies should address unsafe acts, including disciplinary measures for violations. This underscores the seriousness of biosecurity measures and discourages neglect or deliberate breaches.
- **Open Communication**: Creating an environment where employees feel comfortable reporting unsafe acts by colleagues is crucial. This allows immediate action to correct the situation and prevents minor infractions from escalating into major incidents.
- **Leadership Commitment**: Leaders within the institution must demonstrate commitment to biosecurity and safe practices, both in their actions and communication. This helps to instill a culture of safety where unsafe acts are not tolerated.
- **Proactive Analysis**: Regular audits and inspections can identify patterns of unsafe behavior before they result in accidents. Analyzing "near-miss" incidents can also provide valuable insights into potential weaknesses in current safety measures.
- **Psychosocial Support**: Providing support for stress management and maintaining a healthy work environment can reduce the likelihood of unsafe acts borne out of fatigue, stress, or job dissatisfaction.
- **Whistleblower Protections**: Providing safeguards for whistleblowers is key to ensuring that employees feel safe reporting unsafe acts without fear of retaliation.

By proactively addressing the issue of unsafe acts through these measures, laboratories and institutions can significantly enhance their biosecurity and reduce the potential for biosecurity incidents caused by human factors. It is essential to remember that human factors are an integral part of any biosecurity framework and addressing them effectively requires consistent effort and commitment at all levels of the organization.

9.4 LACK OF TEAMWORK

Teamwork is part of the organizational culture and is critical in any scientific setting, particularly in synthetic biology, where collaboration across different domains of expertise is often required. Within the organizational culture, a lack of effective teamwork, characterized by poor communication, ineffective coordination, and minimal collaboration, can significantly escalate the risk of errors, accidents, and biosecurity breaches.

- **Poor Communication**: Misunderstandings or misinterpretations can easily occur when team members do not clearly articulate their thoughts, needs, or concerns. This may lead to errors in experimental design or execution, resulting in accidents or inadvertent breaches of biosecurity protocols. Clear, concise, and frequent communication among team members can mitigate this risk.
- **Ineffective Coordination**: Poor coordination can result in gaps or overlaps in work, misaligned efforts, and inefficient use of resources. This lack of organization can directly compromise safety protocols and increase the likelihood of accidents. Effective coordination can be fostered through regular team meetings, clearly defined roles and responsibilities, and project management tools.
- **Minimal Collaboration**: A lack of collaboration may result in important information not being shared or utilized effectively. It can also prevent the generation of innovative ideas, which often arise from the intersection of different fields and perspectives. Fostering a culture that values and promotes collaboration can enhance problem-solving and decision-making, as well as ensure a comprehensive approach to biosecurity.

Mitigating the risks associated with poor teamwork involves both individual and organizational strategies. Individually, team members can be trained in effective communication, leadership, conflict resolution, and teamwork skills. Organizational strategies might include the creation of a positive and inclusive team culture, the provision of team-building activities, and the implementation of clear procedures for coordination and collaboration.

By strengthening teamwork, laboratories and institutions can not only improve their biosecurity measures but also enhance the overall efficiency and quality of their work in synthetic biology. Thus, promoting effective teamwork should be a key component of any comprehensive approach to managing human factors in biosecurity.

9.5 LACK OF LEADERSHIP

Effective leadership is a crucial element in fostering a culture of safety and biosecurity synthetic biology. Poor leadership, marked by a lack of direction, inadequate support, and insufficient accountability, can exacerbate human errors and increase the risks of accidents or biosecurity breaches.

- **Lack of Direction**: Leaders have the responsibility to provide clear guidance and set explicit goals for their teams. Without direction, team members may not fully understand the protocols or the importance of following them, increasing the risk of biosecurity incidents. Leaders should clarify expectations and ensure team members understand the procedures and the reasons behind them.

- **Inadequate Support**: Leaders need to provide the necessary resources and support for their teams to perform their duties effectively and safely. This includes ensuring access to safety equipment and training, encouraging open communication, and fostering an environment where safety concerns can be raised without fear of reprisal. Without such support, team members may cut corners or avoid following safety protocols, potentially leading to biosecurity breaches.
- **Insufficient Accountability**: Effective leaders should promote accountability by setting an example, holding themselves and their team members responsible for adhering to safety protocols. A lack of accountability can lead to complacency and negligence, with team members feeling that they can bypass protocols without facing any consequences.

To address these issues, a multifaceted approach to leadership development should be pursued. This should involve leadership training programs focusing on communication, team management, and promoting a safety culture. Leaders should be taught to provide clear instructions, offer support, and hold their teams accountable.

Furthermore, institutional policies should ensure that leaders are held accountable for the safety performance of their teams. Such policies can also reinforce the importance of biorisk management, helping to create a culture where these considerations are viewed as critical aspects of all work in synthetic biology.

Effective leadership is a pivotal component in managing the human factors that influence biosecurity in synthetic biology, underscoring the importance of developing and maintaining strong leadership structures within laboratories and research institutions.

9.6 INADEQUATE TRAINING

In synthetic biology, training is an essential component of biosecurity. A lack of adequate training or an incomplete understanding of safety procedures and protocols can significantly increase the risk of errors, accidents, and biosecurity breaches.

- **Understanding of Protocols**: Proper training ensures that individuals working in synthetic biology have a complete understanding of the safety protocols and procedures. Inadequate training may lead to misunderstanding or misapplication of these protocols, leading to accidental releases of organisms or other biosecurity incidents.
- **Safe Laboratory Practices**: Laboratory work in synthetic biology often involves handling potentially hazardous biological materials. Without adequate training in safe laboratory practices, individuals may inadvertently expose themselves or others to risks, or cause a breach in biosecurity.
- **Emergency Response**: Training should also cover emergency response procedures. In the event of an accident or a breach in biosecurity, individuals need to know how to respond effectively to contain the situation and prevent further harm.

- **Ethical and Regulatory Considerations**: Training should not just cover technical skills but also ethical and regulatory considerations. Individuals working in synthetic biology need to understand the ethical implications of their work and the regulations that govern it.

To address these issues, institutions need to implement comprehensive training programs that cover all aspects of biosecurity, from basic laboratory safety to emergency response procedures, ethical considerations, and regulatory compliance.

Training should be an ongoing process, with regular refresher courses to ensure that knowledge and skills are kept up to date. It should also be tailored to the specific needs and risks of the work being conducted. For example, someone working with genetically engineered pathogens would require different training than those working with non-pathogenic organisms.

Furthermore, institutions should monitor and assess the effectiveness of their training programs, for example, through testing and evaluation or by tracking incidents that occur after training has been completed. This can help to identify any gaps or weaknesses in the training and make necessary adjustments.

Adequate training is a fundamental aspect of managing the human factors that contribute to biosecurity in synthetic biology, highlighting the importance of comprehensive, ongoing, and effective training programs in this field.

9.7 LACK OF RESOURCES

Resources – time, materials, equipment, and even personnel – are critical to maintaining a robust biosecurity environment in synthetic biology. A lack of these resources can compromise safety and security, increasing the potential for errors, accidents, and breaches in biosecurity.

- **Time**: Adequate time is necessary to ensure tasks are performed thoroughly and carefully. Rushing procedures due to time constraints can lead to oversights, incomplete work, and increased potential for mistakes. Also, time is required for proper training and refresher courses, as well as for routine safety checks and maintenance.
- **Materials**: Proper materials are essential for safe laboratory work. This includes not only the raw materials used in synthetic biology, but also safety equipment like lab coats, gloves, and goggles. A lack of necessary materials can lead to improvised solutions that may not meet safety standards.
- **Equipment**: Synthetic biology often requires specialized equipment, and a lack of such equipment can lead to suboptimal or unsafe practices. It is crucial to ensure that laboratories are adequately equipped and that all equipment is maintained in good working order. This includes laboratory instruments used in experiments and safety equipment like fume hoods and biosafety cabinets.
- **Personnel**: A lack of sufficient personnel can increase workloads, potentially leading to fatigue, stress, and mistakes. It can also make it challenging to maintain appropriate safety protocols, such as the "two-person rule" for work with specific high-risk pathogens.

To mitigate these challenges, it is necessary to ensure that adequate resources are allocated to biosecurity. This includes budgeting for the procurement and maintenance of necessary materials and equipment, as well as for the hiring and training of sufficient personnel.

Furthermore, resource allocation should consider the need for redundancy and contingency planning. This ensures that operations can continue safely even in the face of unexpected challenges, such as equipment failure or personnel absence.

Institutions also need to prioritize tasks and manage resources effectively. Regular audits and risk assessments can help to identify areas where resources are lacking and prioritize investments in biosecurity. Regular reviews can also help to ensure that resource allocation keeps pace with evolving needs and challenges in synthetic biology.

9.8 LACK OF STANDARDS

Standards are a vital foundation for consistency and quality in any field, including synthetic biology. They provide a common language and understanding, allowing different individuals, teams, and organizations to work together effectively and safely. A lack of clear, consistent, and widely accepted standards can lead to confusion, inefficiency, and a higher risk of errors and accidents.

- **Confusion and Misinterpretation**: Without clear standards, different individuals and teams might adopt different methods and interpretations, leading to inconsistency and confusion. For instance, if there are no standard protocols for handling certain biological materials, different lab workers might adopt different procedures, increasing the risk of mishandling and accidents.
- **Lack of Quality Control**: Standards are critical for maintaining the quality of work. They serve as benchmarks against which performance can be measured. A lack of standards can make it difficult to assess the quality of work or to identify problems and areas for improvement.
- **Hindrances to Collaboration**: In the global field of synthetic biology, teams from different institutions and even different countries often need to collaborate. A lack of shared standards can impede this collaboration, as each group might have its practices and protocols.
- **Impediments to Training and Education**: Standards form the basis for training and education. Without them, it can be challenging to develop effective training programs or ensure that all personnel have the necessary skills and knowledge.

To overcome these challenges, it is necessary to develop and adopt clear, comprehensive, and practical standards for synthetic biology. These standards should cover all aspects of the field, from laboratory procedures to data handling and reporting.

Development of these standards should involve a wide range of stakeholders, including scientists, safety experts, ethicists, and regulators. This ensures that the standards are practical, ethically sound, and compliant with regulatory requirements.

Furthermore, standards should be regularly reviewed and updated to keep pace with the rapid advancements in synthetic biology. This ensures they remain relevant and effective in the face of new technologies and challenges.

Adherence to these standards should also be encouraged and enforced through measures such as regular audits, certification schemes, and incentives. Training and education should also emphasize the importance of standards and provide practical guidance on implementing them.

9.9 STRESS

Stress is an inevitable part of any work environment, including synthetic biology. While a certain amount of stress can stimulate productivity and creativity, high stress levels can negatively impact performance, decision-making abilities, and overall well-being, increasing the risk of mistakes and accidents.

- **Impaired Cognitive Performance**: High stress levels can impair cognitive functions such as attention, memory, and decision-making. In a synthetic biology lab, this could lead to errors in experimental procedures, data analysis, or safety protocols, potentially resulting in harmful incidents.
- **Reduced Physical Performance**: Stress can also impact physical health and performance. It can cause fatigue, impair motor coordination, and reduce immune response, making individuals more susceptible to accidents and illnesses.
- **Emotional Disturbance**: High stress levels can lead to emotional disturbances such as irritability, anxiety, and depression. These emotional states can affect interpersonal relations, team dynamics, and overall workplace morale, further exacerbating stress levels and risks of mistakes.
- **Burnout**: Chronic, unmanaged stress can lead to burnout, which is a state of emotional, mental, and physical exhaustion. Burnout can lead to disengagement, reduced productivity, and increased error rates, posing significant risks to biosecurity.

To manage the impacts of stress on biosecurity, several strategies can be employed:

- **Workload Management**: Overwhelming workload is a major contributor to workplace stress. It is important to ensure that workloads are manageable and that individuals have enough time to complete their tasks without feeling rushed or overwhelmed.
- **Work-Life Balance**: Encouraging a healthy work-life balance can help reduce stress. This might involve flexible work hours, sufficient time off, and support for personal or family needs.
- **Supportive Work Environment**: A supportive work environment, with open communication, positive interpersonal relations, and opportunities for professional growth, can help mitigate stress.

- **Stress Management Training**: Providing training on stress management techniques, such as relaxation exercises, mindfulness, and time management, can equip individuals with the skills to manage stress effectively.
- **Resources**: Providing resources for mental and emotional health support, such as counseling services or employee assistance programs, can also be beneficial.

Recognizing and managing stress is a critical part of biosecurity in synthetic biology. By taking proactive steps to reduce workplace stress and support employee well-being, we can create a safer, more productive work environment.

9.10 FATIGUE

Fatigue is a state of chronic tiredness or exhaustion, which can be physical, mental, or both. It is a significant human factor to consider in biosecurity in synthetic biology, as it can lead to a decrease in vigilance, attention, and decision-making abilities, thus increasing the risk of accidents and errors.

- **Physical Fatigue**: Physical fatigue can result from a heavy workload, insufficient rest, or poor health. It can affect manual dexterity and coordination, slowing reaction times and increasing the risk of accidents in a lab setting.
- **Cognitive Fatigue**: Cognitive fatigue, also known as mental fatigue, can arise from prolonged periods of cognitive activity, high stress levels, or inadequate sleep. It can impair important cognitive functions such as attention, memory, and decision-making, potentially leading to mistakes in experimental procedures, data analysis, or safety protocol compliance.
- **Effects on Performance**: Fatigue can lead to reduced productivity, lower quality of work, and an increased propensity for risk-taking behavior. In a synthetic biology laboratory, this might result in errors, security breaches, or safety incidents.

To manage the risks associated with fatigue, it is essential to implement several strategies:

- **Workload Management**: Assigning reasonable workloads and ensuring adequate rest periods can help to prevent excessive fatigue. This might involve implementing shift work schedules, setting maximum work hours, or encouraging regular breaks.
- **Healthy Work Environment**: A work environment that promotes health and well-being can help to mitigate fatigue. This includes proper lighting, comfortable temperatures, ergonomic workstations, and access to healthy meals and hydration.
- **Sleep and Rest Policies**: Encouraging adequate sleep and rest can help to prevent fatigue. This might involve educating employees about sleep hygiene, providing facilities for rest during breaks, or implementing policies that discourage excessive overtime.

- **Health and Wellness Programs**: Programs that promote physical activity, stress management, and overall health and wellness can help to combat fatigue. These might include access to fitness facilities, wellness resources, or health screenings.
- **Training and Awareness**: Training employees on the signs of fatigue, the risks associated with fatigue, and strategies for managing fatigue can help to raise awareness and promote proactive management of fatigue.

Recognizing and managing fatigue is a vital part of maintaining biosecurity in synthetic biology. By adopting comprehensive fatigue management strategies, laboratories can enhance safety, productivity, and employee well-being.

9.11 HUMAN ERROR

Human error refers to actions or decisions that produce unintended or unexpected results, often leading to negative outcomes or accidents. It is an inevitable aspect of human activity and a significant concern in biosecurity for synthetic biology. The consequences of human error in this field can range from minor procedural discrepancies to severe biosecurity breaches with potential health and environmental implications.

- **Types of Human Error**: Errors can be broadly categorized into two types – mistakes and slips. Mistakes occur when actions go as planned but fail to achieve the intended outcome, which is often due to a lack of knowledge or incorrect problem-solving strategy. Slips, on the other hand, are actions that do not go as planned, usually because of inattention or misplaced focus.
- **Contributing Factors**: Several factors can increase the likelihood of human error. These include a lack of knowledge or skill, time pressure, physical and mental fatigue, high stress levels, inadequate resources, poor communication, and ineffective supervision or leadership.

To manage the risks associated with human error, a proactive and systematic approach is needed:

- **Training and Education**: Regular training and education programs can ensure that all individuals possess the necessary knowledge and skills to perform their tasks correctly. It also helps keep them updated with the latest protocols and safety guidelines.
- **Error Reporting and Analysis**: Establishing a system for error reporting and analysis can help in identifying trends, understanding root causes, and implementing preventive measures. A blame-free reporting culture encourages individuals to report their mistakes without fear of punishment, providing valuable learning opportunities.
- **Standard Operating Procedures**: Clearly written and easily accessible standard operating procedures (SOPs) provide a reference point for correct

practice and help to minimize errors. Regular audits and updates ensure SOPs stay current and relevant.

- **Workplace Design and Automation**: A well-designed workspace can minimize the potential for error. This includes ensuring that equipment is ergonomically designed and that the workplace layout is logical and intuitive. Automation of routine or complex tasks can also reduce the potential for human error.
- **Stress Management and Wellness Programs**: Implementing programs that promote stress management, physical health, and mental wellness can help in mitigating factors like fatigue and stress that contribute to human error.

Human error is a significant challenge in maintaining biosecurity in synthetic biology. Recognizing this, understanding its causes, and implementing strategies to minimize its occurrence can help to maintain high safety and security standards.

9.12 INDIVIDUAL DIFFERENCES

Synthetic biology, like any other scientific discipline, involves a diverse range of individuals, each with unique characteristics such as age, experience, personality, and physical abilities. These individual differences can have a significant impact on safety and contribute to accidents in several ways:

- **Experience and Skill Level**: Individuals with varying levels of experience and skill may approach tasks differently. Novice workers may lack the experience to foresee potential issues, while experienced workers might become complacent or overconfident, leading to oversights.
- **Age**: In synthetic biology, the human factor of age plays a significant role in biosecurity considerations. A range of cognitive, physical, and psychological abilities integral to upholding biosecurity standards are influenced by age. Cognitive abilities such as problem-solving, decision-making, memory, and processing speed are all vital for tasks like accurately adhering to protocols, identifying potential biosecurity risks, and dealing with unexpected circumstances. Such abilities have been shown to evolve with age. Experience, often associated with advanced age, can bolster decision-making and problem-solving capabilities. Certain cognitive abilities like memory and processing speed may experience a decline with increasing age, although this is at the individual level. Consequently, an imbalance between task demands and cognitive abilities due to varying ages could escalate the risk of human error in biosecurity.

  Physical abilities encompassing strength, dexterity, visual acuity, and hearing can also influence an individual's capacity to carry out laboratory tasks securely and efficiently. With age, these abilities might decline, potentially impacting the secure handling of biological materials or the proper operation of lab equipment.

  On the psychological front, factors such as stress tolerance and motivation can also demonstrate variation with age. While older workers may showcase superior stress management skills from their life experience, their motivations and attitudes toward work might differ from those of younger employees. This may influence their adherence to biosecurity protocols.

Addressing these age-related factors requires designing biosecurity training and protocols considering the broad spectrum of cognitive, physical, and psychological capabilities. Regular assessments and refresher courses can aid in maintaining high biosecurity standards among employees of all ages. Moreover, making necessary accommodations can ensure that all workers, irrespective of their age, can safely and effectively perform their duties. It is also vital to cultivate an inclusive workplace culture that appreciates contributions from all age groups and leverages the unique strengths of younger and older employees.

- **Personality Traits**: Certain personality traits can affect how individuals approach their work. For example, those with a higher tolerance for risk may be more likely to bypass safety protocols. In contrast, individuals with high levels of conscientiousness are more likely to follow procedures carefully.
- **Physical Abilities**: Differences in physical abilities can also influence safety. For example, individuals with certain physical limitations (such as strength in the hands or height) might find it challenging to handle lab equipment effectively, leading to potential errors or accidents.
- **Mental Health – Neurotypical vs. Neurodivergent**: Mental health and cognitive diversity are integral considerations for the functioning of any work environment, including synthetic biology laboratories. Neurotypical individuals, whose neurological development and functioning are within the standard norms, might process information, interact, and react to their environment differently than neurodivergent individuals, who have neurological variations such as autism, ADHD, and dyslexia.

Neurodivergent individuals often exhibit unique strengths, such as a heightened focus on tasks of interest, innovative thinking, or heightened creativity. These traits can be highly advantageous in a research setting, leading to unique insights and solutions. However, they may also face unique challenges related to their neurodivergence, such as social communication challenges, sensory issues, or difficulty with organizational skills. These differences can affect their interaction with their work environment and potentially influence the risk of human error in biosecurity.

For instance, someone with ADHD might struggle with focusing on repetitive tasks or following lengthy protocols, potentially increasing the chance of error. On the other hand, an individual on the autism spectrum might excel in attention to detail and routine tasks but might struggle with unexpected changes in protocols or social-communicative aspects of teamwork.

Supporting neurodivergent individuals in the lab goes beyond just providing accommodations. It involves creating an inclusive work environment that values cognitive diversity and leverages everyone's strengths. This might involve offering flexible work procedures, providing clear and concise instructions, allowing the use of assistive technologies, or providing training to all staff about neurodiversity. It is important to note that this is a new area, and everyone may be most knowledgeable about their specific strengths and needs for accommodation.

In terms of mental health, it is crucial to maintain an environment that reduces unnecessary stressors, as high stress can exacerbate conditions and impair performance. Furthermore, promoting a supportive and understanding workplace culture can enhance overall mental well-being, which is important for maintaining high biosecurity standards. Therefore, considering neurotypical and neurodivergent individuals' needs and strengths is crucial for ensuring high biosecurity standards and an inclusive, efficient, and productive work environment.

- **Cultural Backgrounds/Nationality**: Culture and nationality play pivotal roles in shaping individuals' identities and behaviors, influencing their modes of communication, values, beliefs, and thought processes. Such differences become significant in synthetic biology, where these variations in human factors can impact biosecurity. Language and communication styles can vary across cultures. Misinterpretations or miscommunications can lead to errors in a biosecurity setting where precision in communication is vital. Moreover, cultural differences often dictate attitudes toward authority and hierarchy. In some cultures, it may be viewed as disrespectful to question authority or deviate from prescribed protocols. By fostering an environment that encourages open dialogue and feedback across all hierarchical levels, a culture of safety can be created, enabling quick identification and resolution of potential biosecurity risks. Cultural background can also influence risk perception and safety attitudes. Acknowledging these cultural differences can lead to the development of more effective training and safety protocols tailored to the cultural context of lab personnel.

Furthermore, unique safety cultures are present at every level – from individual lab groups to departments and the institution. It is crucial to ensure that these cultures are in alignment with each other. For instance, if an institution fosters a robust safety culture that values openness and inclusivity, but a particular lab group does not prioritize safety, this discrepancy may result in an increased number of safety and security incidents within that group. Such negative safety cultures can propagate quickly due to human factors, as detailed in a discussion in Chapter 9. Recognizing these differences and having strong leadership to synchronize these varying cultures is essential. Addressing concerns promptly and efficiently plays a significant role in preserving a culture of biorisk management. We are only as strong as our weakest link.

To mitigate these risks, the following strategies should be considered:

- **Comprehensive Training and Continuous Learning**: Providing comprehensive training tailored to the skill level of each individual is critical. For new employees, this might involve extensive onboarding, while for experienced employees, this could mean refresher courses or advanced training.
- **Accommodation of Physical Abilities**: Work environments should be inclusive and accessible for individuals of all physical abilities. Ergonomic designs can enhance comfort and efficiency, reducing the likelihood of accidents.

- **Consideration of Personality Traits**: Understanding the personality traits of team members can enable the creation of balanced teams and a working environment that acknowledges these differences. For instance, risk-tolerant individuals could be paired with more cautious ones to ensure a balance.
- **Age-Appropriate Workload and Tasks**: Recognizing the strengths and limitations at different ages can allow the assignment of age-appropriate tasks and ensure a balanced workload.

Understanding and accounting for individual differences can thus enhance safety, reduce accidents, and create a more effective, inclusive work environment in synthetic biology. By identifying and addressing these factors, it is possible to improve safety and reduce the risk of accidents and incidents in the workplace.

9.13 MENTAL/EMOTIONAL HEALTH AND INSIDER THREAT TO BIOSECURITY BEST PRACTICES

Mental and emotional health are an often overlooked but crucial aspects of biosecurity, particularly in synthetic biology. As synthetic biology research becomes more accessible and widespread, the potential for insider threats – risks posed by individuals within an organization with access to sensitive information or materials – increases. Mental and emotional health issues can exacerbate these risks.

Stress, burnout, and other mental health issues can affect a researcher's judgment and performance, potentially leading to biorisk management lapses. For instance, a stressed researcher might forget to follow a critical safety protocol, or a researcher dealing with mental or emotional health issues might become disgruntled and intentionally misuse their access to sensitive biological materials or data.

Moreover, the increasing accessibility of synthetic biology tools and techniques means that individuals outside traditional research environments, such as DIY biologists, also need to be aware of the mental health aspects of biosecurity. Without the support structures typically found in institutional settings, these individuals might be more vulnerable to the negative effects of stress and other mental and emotional health issues on their work.

To address these challenges, a multifaceted approach is needed. The following are some strategies.

9.13.1 Promoting a Healthy Work Culture

Creating a work environment that values mental and emotional health is crucial. This includes fostering a supportive culture, providing resources for stress management and support, and encouraging work-life balance.

Creating a work environment that values mental and emotional health is essential for the well-being and productivity of employees, as well as for minimizing potential biosecurity risks. Here are three steps institutions can take to promote mental and emotional health.

9.13.1.1 Implement Mental and Emotional Health Programs

Institutions can establish mental and emotional health programs that provide resources for employees to manage stress, anxiety, and other issues. This could include counseling services, mindfulness training, stress management workshops, or access to professionals. In addition, mental and emotional health awareness campaigns can help reduce stigma and encourage individuals to seek help when needed.

Creating a work environment that values mental and emotional health is particularly important in a synthetic biology lab, where high-stress situations can be common and the consequences of mistakes can be significant. Here are three steps specifically tailored for a synthetic biology lab setting:

1. **Establish a Lab Safety and Well-being Committee**: This committee could oversee the mental well-being of lab members and their physical safety. The committee can regularly check in with lab members, provide a platform for discussing stressors and challenges, and develop strategies to promote mental and emotional health and stress management in the lab.
2. **Provide Training on Stress Management in Lab Settings**: In a synthetic biology lab, stress might arise from various sources, such as the pressure to produce results, the handling of sensitive or dangerous biological materials, or the need to stay up to date with rapidly evolving techniques and technologies. Providing tailored training on managing these specific stressors can help lab members handle them more effectively. This could be part of the regular lab meetings or dedicated workshops.
3. **Promote a Balanced Lab Life**: Encourage lab members to take regular breaks, especially during long experiments. Promote the importance of time off and discourage a culture of working excessive hours. Ensure lab members have time for relaxation and recreational activities. Also, consider organizing regular social activities for the lab team, as these can foster a supportive community and provide a break from the work environment.

A supportive and mentally healthy lab environment does not just benefit the individuals working in the lab – it can also contribute to better science and safer lab practices.

9.13.1.2 Promote Work-Life Balance

A healthy work-life balance can significantly improve mental and emotional health. Institutions can promote this by implementing policies that support flexible work schedules, encourage regular breaks, and respect personal time. For example, limiting after-hours work and email, encouraging time off, and recognizing the importance of personal and family time can all contribute to better work-life balance.

Promoting work-life balance in a synthetic biology lab setting is crucial for maintaining the well-being of lab members and fostering a productive, creative, and safe working environment. Here are three specific steps that can be taken:

1. **Flexible Work Schedules**: Not all lab work needs to be conducted during traditional working hours. Where possible, allow lab members to have some flexibility in their schedules. This might involve letting them start earlier or

finish later to avoid peak commute times or allowing them to take time off during the day for personal appointments or family responsibilities.

2. **Limit After-Hours Work**: While there may be times when work outside of regular hours is necessary, try to limit this as much as possible. Encourage lab members to disconnect from work in their off hours and discourage the sending of work-related emails or messages outside of agreed working hours. If after-hours work is necessary, ensure this is balanced with adequate time for rest.
3. **Encourage Regular Breaks and Time Off**: Regular breaks during the workday can help prevent burnout and improve productivity. Encourage lab members to take short breaks throughout the day, and ensure they take time off for meals. Also, encourage lab members to take their full vacation time, and discourage a culture where people feel they cannot take time off.

9.13.1.3 Foster a Supportive Culture

The culture of an institution plays a significant role in mental and emotional health. Leadership should model and promote a culture of respect, inclusivity, and open communication. This should involve regular check-ins with employees, creating a safe space for employees to share their concerns and experiences, and fostering a collaborative rather than competitive environment. Recognition of employees' efforts and achievements can also contribute to a supportive culture.

Fostering a supportive culture in a synthetic biology lab is crucial for the well-being and productivity of lab members. Here are three steps that can help achieve this:

1. **Promote Open Communication**: Encourage lab members to share their ideas, concerns, and challenges. Regular team meetings can provide a forum for open discussion, while one-on-one check-ins can allow for more personal conversations. Make it clear that all voices are valued and that constructive feedback is welcome.
2. **Recognize and Celebrate Achievements**: Regularly acknowledge the contributions and accomplishments of lab members. This can be as simple as verbal recognition in a team meeting, or more formal recognition like awards or acknowledgments in publications. Celebrating successes, big and small, can boost morale and foster a sense of teamwork.
3. **Provide Support and Resources**: Ensure that lab members have the resources and support they need to do their work effectively. This should include training opportunities, mentorship programs, or resources. Additionally, ensure that lab members know where to go for help when needed, whether it is a technical question about a lab procedure or a personal issue impacting their work.

9.13.2 Training and Awareness

Researchers should be trained not only in technical biorisk management protocols but also in recognizing and managing stress and other mental and emotional health issues. Awareness of the potential impact of mental and emotional health on biosecurity should be promoted at all levels of an organization. This is essential for creating a safe and supportive environment. Here are three steps an institution can take to achieve this:

1. **Comprehensive Training Programs**: Develop comprehensive training programs incorporating technical biorisk management protocols, personality types, strengths, communication styles, and mental and emotional health awareness. These programs should cover topics such as risk assessment, proper handling of hazardous materials, safety procedures, and identifying potential mental and emotional health challenges. Include case studies and interactive sessions to enhance understanding and engagement.
2. **Collaboration with Mental and Emotional Health Professionals**: Collaborate with mental and emotional health professionals or experts to provide specialized training sessions on recognizing and managing stress and other issues. These sessions can focus on stress management techniques, building resilience, recognizing signs of distress in oneself and others, and understanding the impact of mental and emotional health on overall well-being and work performance.
3. **Promote a Supportive Culture**: Create an organizational culture that values mental and emotional well-being and encourages open discussions about it. This includes promoting a stigma-free environment where individuals feel comfortable seeking help and support. Encourage supervisors and mentors to actively listen and engage in supportive conversations with their team members. Provide resources such as access to services, counseling, or employee assistance programs.

Additionally, consider incorporating ongoing professional development opportunities related to mental and emotional health, personality types, and biosecurity, such as seminars, workshops, or guest lectures, to keep researchers updated on emerging best practices and research findings.

By integrating technical biorisk management protocols and mental and emotional well-being and personality types training, institutions can equip researchers with the knowledge and skills to promote a safe, secure, and supportive research environment. This holistic approach supports the well-being of researchers and enhances their ability to recognize and address potential mental and emotional health issues while maintaining strong biosecurity practices.

9.13.3 Screening and Monitoring

Regular mental and emotional health well-being screenings and monitoring can help identify individuals who might be struggling and provide them with the support they need. However, such measures need to be implemented with care to respect privacy rights and avoid stigmatizing mental and emotional health issues.

Developing regular mental and emotional well-being screenings and monitoring processes within an institution requires a thoughtful and balanced approach that respects privacy rights and avoids stigmatization. Here are three ways an institution can achieve this:

1. **Establish Confidential and Voluntary Programs**: Create confidential and voluntary mental and emotional health screening programs that allow individuals to participate based on their discretion. Ensure that participation is optional and communicate that the purpose is to support individuals in maintaining their mental well-being. Emphasize that the information collected will be treated confidentially and used solely to provide appropriate support.
2. **Engage Trained Professionals**: Involve coaches, mentors, or trained staff who can conduct the screenings and provide support. These professionals should adhere to strict confidentiality guidelines and ethical standards. They can administer assessments or interviews to identify signs of distress, anxiety, depression, or other mental and emotional health issues. The focus should be on early detection and providing appropriate resources rather than diagnosing or labeling individuals.
3. **Ensure Privacy and Confidentiality**: Safeguard privacy rights by implementing robust policies and procedures to protect individuals' personal information. Ensure that the collected data is securely stored, accessible only to authorized personnel, and used solely for providing support and resources. Establish clear guidelines on how the information will be managed, shared, and anonymized to avoid any potential breach of privacy. Communicate these policies transparently to participants to build trust in the screening process.

In addition to regular screenings, institutions should prioritize creating a supportive environment that encourages open conversations about mental and emotional well-being. This can be achieved by offering resources such as mental and emotional health support services, employee assistance programs, and training programs that promote mental and emotional well-being and stress management.

Remember, it is crucial to balance the goals of supporting individuals' mental and emotional health with their privacy rights and avoiding stigmatization. Open dialogue, education, and support systems play a vital role in fostering a culture of well-being within the institution.

9.13.4 Having Clear Policies and Procedures

Organizations should have clear policies and procedures for managing situations where an individual's mental and emotional health might pose a biosecurity risk. This includes procedures for reporting concerns and taking appropriate action.

Creating clear organizational policies and procedures for managing situations where an individual's mental and emotional health might pose a biosecurity risk is essential for maintaining a safe and secure working environment. Here are three important points to consider:

1. **Establish Reporting Mechanisms**: Develop transparent and confidential procedures for reporting concerns about an individual's general well-being and its potential impact on biosecurity. Create multiple channels for reporting, such as designated individuals or anonymous reporting systems, to ensure that anyone can raise concerns without fear of reprisal. Communicate these reporting mechanisms to all organization members and provide training on how to utilize them effectively.
2. **Assessments and Evaluation**: Establish a process for conducting assessments and evaluations when concerns are raised about an individual's mental and emotional health and its potential impact on biosecurity. This might involve forming a specialized committee or team responsible for evaluating the situation, including mental health professionals, biorisk officers, human resources representatives, and relevant stakeholders. The assessment should be conducted with empathy, confidentiality, and adherence to privacy rights.
3. **Action and Support**: Develop guidelines for taking appropriate action based on the assessment outcomes. This may include providing support and resources to the individual such as access to mental health services, counseling, or employee assistance programs. Additionally, establish protocols for reassignment of duties, temporary leave, or other accommodations that may be necessary to ensure both the individual's well-being and the overall biosecurity of the organization. Ensure these actions are implemented with sensitivity, confidentiality, and respect for the individual's rights.

It is crucial to strike a balance between addressing potential risks to biosecurity and safeguarding the rights and privacy of the individual involved. These policies and procedures should be communicated clearly throughout the organization, with training provided to relevant staff members to ensure consistent implementation.

Regular reviews and updates of these policies and procedures are necessary to adapt to changing circumstances, evolving best practices, and legal requirements.

By addressing the mental and emotional health aspects of biosecurity in synthetic biology, we can help ensure that this exciting field advances in a way that is safe, secure, and respectful of the well-being of all involved.

9.13.5 Considering the "Glass Ceiling" Factor

The term "glass ceiling" refers to an invisible barrier that prevents certain groups, particularly women and racial or ethnic minorities, from advancing to higher positions in corporate hierarchies or research institutions, regardless of their qualifications or achievements. This systemic barrier can significantly impact the career trajectory and well-being of these individuals and can influence their willingness and motivation to comply with institutional rules and regulations.

- **Impact on Well-Being**: Experiencing the glass ceiling can lead to various negative outcomes for individuals. It can cause significant stress, frustration, and dissatisfaction, as hard work and potential are not rewarded or recognized. This can also lead to a sense of isolation and alienation, as individuals feel undervalued and underrepresented within the organization. Over time, these experiences can impact mental health, resulting in increased rates of anxiety, depression, and burnout. Beyond the individual, the glass ceiling can also impact the overall workplace environment, leading to lower morale and reduced team cohesion. It can perpetuate a culture of exclusion and inequality, where diverse perspectives and contributions are not valued or included.

- **Impact on Compliance**: The impact of the glass ceiling extends beyond individual well-being and can also affect compliance with rules and regulations within an organization. Employees who feel undervalued or discriminated against may become demotivated and disengaged, resulting in a lackadaisical attitude toward following institutional policies and procedures. In a research or laboratory setting, this could have severe implications for biorisk management. If employees feel they are not being treated fairly or their contributions are not recognized, they may be less inclined to adhere to biorisk management protocols. This jeopardizes not only their safety but also that of their colleagues and the broader community.

To effectively address the glass ceiling phenomenon, organizations need to cultivate an inclusive and fair work culture. Three steps that organizations can take to prevent or reduce the glass ceiling effect and create a better sense of belonging, inclusion, and opportunities for their employees are:

- **Establish Clear Pathways for Advancement**: Organizations should establish clear and transparent pathways for career advancement, ensuring that employees have equal opportunities for growth and development. This involves implementing performance evaluation systems that are fair, objective, and free from biases. Additionally, organizations can create mentorship or career development programs that provide guidance and support to individuals who may face unique challenges due to the glass ceiling effect. By promoting a culture of fairness and equal opportunities, organizations can empower employees to reach their full potential and mitigate the impact of the glass ceiling.
- **Implement Equal Pay Policies**: Addressing wage gaps is crucial for dismantling the glass ceiling. Organizations should implement policies that ensure equal pay for equal work, regardless of gender, race, ethnicity, or any other protected status. Regular audits should be conducted to ensure compliance and identify areas of improvement. By taking a strong stance on pay equity, organizations foster a sense of fairness and respect, which can motivate all employees to excel in their roles.
- **Incorporate "Cognitive Diversity" for Leadership Selection**: Cognitive diversity, referring to differences in how people process information and perspectives, can enhance team performance when tackling complex challenges. This has powerful implications for increasing diversity within teams and shattering the glass ceiling. Organizations that select leaders based on cognitive factors rather than demographic traits see greater diversity emerge in

their upper ranks. Often, teams lack diversity, dominated by similar thinking styles and approaches to problem-solving. Expanding demographic diversity alone has not shattered the glass ceiling. However, actively fostering cognitive diversity within teams will lead to better outcomes. Team members who think differently, process information in varied ways, and offer alternative perspectives drive innovation, avoid groupthink, and make better decisions. Organizations that construct teams based on cognitive factors rather than demographic traits see greater diversity emerge in group composition and performance.[162] Therefore, making cognitive diversity an explicit criterion for team and leadership selection and development, rather than focusing on surface-level traits, can enhance diversity while improving performance.

These steps, when implemented collectively, can help organizations create a more inclusive and equitable work environment where employees feel valued, supported, and motivated to excel. By breaking down the barriers imposed by the glass ceiling, organizations can harness the full potential of their diverse workforce, leading to improved productivity, innovation, and overall organizational success.

9.14 CONCLUDING THOUGHTS

Addressing human factors in biosecurity is essential for effectively managing risks and enhancing biosecurity in the rapidly evolving field of synthetic biology. Integrating human behavior and organizational culture into biosecurity strategies can lead to a more resilient, responsive, and practical approach to handling the complexities of this field. Human elements play a critical role in maintaining biosecurity, emphasizing the need for a comprehensive and human-centric approach. The complexities of managing human aspects, such as communication breakdowns, complacency, leadership, and training deficiencies, must be thoroughly considered in any biosecurity plan. This chapter underscores the necessity of integrating human factors as a fundamental aspect of effective biosecurity management in synthetic biology. Technological and procedural measures alone are not sufficient when considering the human element. The focus on fostering a biosecurity-conscious mindset among researchers, effective communication, leadership engagement, and comprehensive training reflects a deep understanding of the dynamics of laboratory environments. By addressing these human factors, synthetic biology research can progress in a safer, more secure, and responsible manner, ensuring that the field develops with a keen awareness of the potential risks and the means to mitigate them effectively.

9.15 KEY TAKEAWAYS

- **Importance of Human Factors**: Emphasizes the critical role of human behavior and organizational culture in biosecurity.
- **Communication Breakdowns**: Highlight the impact of miscommunication in biosecurity and the need for effective communication strategies.
- **Complacency Risks**: Identify complacency as a significant biosecurity risk and underscore the need for constant vigilance.

- **Leadership and Training**: Stress the importance of leadership engagement and comprehensive training programs in promoting biosecurity.
- **Policy Development and Reporting**: Advocate for robust policy development and efficient reporting systems for biosecurity incidents.
- **Comprehensive Approach**: Calls for a comprehensive approach to biosecurity that incorporates all human factors.

9.16 THOUGHT-PROVOKING QUESTIONS

9.1 Researchers/scientists play a part in a more extensive system where responsibility is shared among many parties, including government regulators, industry leaders, and the public in ensuring biosecurity in synthetic biology to prevent potential biosecurity threats. What responsibilities should they hold in this role?

9.2 What would be some challenges in updating the current biorisk management guidelines to include "sociotechnical systems?" In such systems, a system's technical and social aspects are intertwined and can influence each other. In the context of biosecurity, can this approach help to identify potential vulnerabilities that might not be apparent when looking at the technical aspects alone? How might this be an effective strategy?

9.3 Which biosecurity incidents from the past can be traced to the inadequate incorporation of human factors? What lessons can we learn from this scenario? Have those inadequacies been addressed in newer versions of biorisk management guidelines?

9.4 What would you include in a hypothetical scenario where a failure to incorporate human factors leads to a biosecurity breach in synthetic biology? What lessons can we learn from this scenario?

9.5 What does modeling the integration of human factors into biosecurity in a real-world laboratory setting look like? What paraments should be included in the model?

9.6 What aspects of human behavior and decision-making should be further understood to enhance biosecurity measures in synthetic biology?

9.7 What systemic factors influencing biosecurity must be formally addressed in official policies and guidelines for a proper risk assessment?

9.8 What policies can be developed to ensure human factors are considered in biosecurity for synthetic biology?

9.9 How can organizations foster a culture of biosecurity that considers human factors?

9.10 How might future developments in synthetic biology challenge our current understanding of human factors in biosecurity?

10 Implementation of Biosecurity Measures in Different Settings

Biosecurity measures should be implemented across various settings where biological research and applications occur. This chapter underscores the diverse environments in which biosecurity is a critical concern, ranging from high-level research laboratories to educational institutions, healthcare facilities, industrial settings, start-up companies, and DIY laboratories. The chapter highlights the importance of robust protocols across all settings to prevent accidental release of biological materials and to ensure the safety of lab workers, scientists, students, and the public. The chapter emphasizes the importance of a proactive and informed approach to biosecurity. It advocates for continuous risk assessment, regular training and awareness programs, and developing a culture of biosecurity and responsibility at all levels. This comprehensive approach is crucial for ensuring effective biosecurity across diverse settings, thereby protecting public health, economic interests, and environmental integrity.

DOI: 10.1201/9781003423171-10

10.1 ACADEMIC SETTING

Implementing biosecurity measures in an academic setting is vital for the safety of students, faculty, and staff, as well as for the wider community and environment. The strategies adopted may differ between large research universities and small colleges, due to the scope and nature of the work conducted, resources available, and the level of risk involved.

10.1.1 Large Research Universities

Large research universities often conduct high-level research, including synthetic biology research involving potentially biohazardous material. Therefore, their biosecurity measures need to be comprehensive:

- **Institutional Biorisk Committee (IBC)**: Universities should have an IBC to oversee and review all research involving recombinant or synthetic nucleic acid molecules and other biohazardous material.
- **Training Programs**: Universities should provide comprehensive biorisk management training to all laboratory personnel. Training should include general biosafety practices and specific procedures for handling synthetic organisms.
- **Facilities and Equipment**: Research universities should ensure that appropriate facilities and equipment, such as biosafety cabinets and autoclaves, are available and maintained. Laboratories should be designed for containment, with features such as self-closing doors and negative air pressure systems.
- **Emergency Response Plans**: Given the high level of research, these universities should have detailed emergency response plans, including procedures for responding to spills or exposures, reporting incidents, and coordinating with external emergency responders.

An added risk factor at large research universities is a research culture where students and postdoctoral fellows are expected to work extreme hours with little to no time for rest. The competitive nature of academia, where the number of publications and the prestige of the journals where the work is published become the currency for graduation, obtaining a job, and tenure and promotion, creates a risky environment where scientists are stressed and exhausted continuously. This can lead to accidents, incidents, insider threats, and the development or exacerbation of mental and emotional health problems. Creating a culture that allows for rest and relaxation is an important part of the scientific endeavor. University administrators should develop policies and set a positive tone from the top to effect change.

10.1.2 Small Colleges

At smaller colleges, research involving biohazardous materials may be less common and often less high-risk, but basic biosecurity measures are still important:

- **Biorisk Officer/Committee**: If an IBC is not required or feasible, a biosafety officer or a smaller committee can be assigned to oversee biorisk management practices, review proposed research for potential risks, and provide guidance to faculty and students.

- **Training**: Even if the risks are lower, basic biorisk management training should still be provided to all students and faculty working in labs. This should cover basic laboratory safety practices, procedures for handling lower-risk biohazardous materials, and steps to take in the event of an accident.
- **Basic Safety Equipment**: Basic safety equipment, such as lab coats, gloves, and safety glasses, should be available and used appropriately. If any higher-risk work is being conducted, appropriate containment equipment should be provided.
- **Safety Procedures**: Clear safety procedures should be in place, including waste disposal, equipment decontamination, and incident reporting.

In both large research universities and small colleges, fostering a culture of safety and responsibility is key. Everyone should understand the potential risks and their role in minimizing those risks, and there should be clear channels for raising concerns and reporting incidents.

10.2 PHARMACEUTICAL INDUSTRY

Biosecurity measures in the pharmaceutical industry play a vital role in protecting personnel, the environment, and the integrity of the drug development process. Depending on the size of the company and the nature of their work, these measures can take various forms.

10.2.1 Startups

Startups often operate on a smaller scale and may have limited resources, but biosecurity remains crucial:

- **Risk Assessment**: Even at a small scale, startups should conduct thorough risk assessments for all work involving potentially biohazardous materials. This can help to identify necessary safety measures and procedures.
- **Training**: Startups should ensure that all staff receive appropriate biorisk management training. This could be provided in-house, or staff could be sent on external training courses.
- **Safety Equipment and Facilities**: Even if a startup operates in a smaller or shared space, appropriate safety equipment and containment measures should be in place. This should include personal protective equipment (PPE), biological safety cabinets, and waste disposal systems.

10.2.2 Large-Scale Manufacturers

Large pharmaceutical companies often work with a broad range of biohazardous materials on a large scale, and so their biosecurity measures need to be extensive:

- **Strict Access Controls**: Due to the nature of large scale and volume, manufacturers need to have rigorous controls on who can access areas where biohazardous materials are present. This can include security clearance

requirements, escorts for visitors, and strict logging of all personnel entering and exiting these areas.
- **Institutional Biorisk Committee (IBC)**: Large companies should have an IBC to oversee and review all work involving potentially biohazardous materials.
- **Comprehensive Training Programs**: Large manufacturers should provide comprehensive biorisk management training programs to all personnel including the unique risks posed by large equipment and volumes.
- **Advanced Safety Equipment and Facilities**: Large manufacturers should have advanced and specialized safety and containment facilities, including state-of-the-art biosafety cabinets, cleanrooms, and waste disposal systems, and a strict maintenance engineering program.
- **Emergency Response Plans**: Given the scale of their operations, large manufacturers should have detailed emergency response plans and dedicated response teams.

10.2.3 Vaccine Manufacturers

Biosecurity in vaccine development is particularly important, as it often involves work with live pathogens:

- **Strict Access Controls**: Due to the nature of vaccine development research, it is important to have strict controls on who can access areas where biohazardous materials are present. This can include security clearance requirements, escorts for visitors, and strict logging of all personnel entering and exiting these areas.
- **Specialized Containment Facilities**: Vaccine development often requires specialized containment facilities, such as those classified as high containment.
- **Virus/bacteria Handling Procedures**: Special procedures are needed for handling viruses/bacteria, including specific containment measures, inactivation procedures, and waste disposal methods.
- **Vaccine Testing Procedures**: Safety and biosecurity measures should be in place for all stages of vaccine testing, including preclinical animal testing and clinical trials in humans.

In all these settings, a commitment to biosecurity at all levels of the organization is essential. This includes fostering a culture of safety, providing appropriate training, and ensuring compliance with all relevant regulations and guidelines.

10.3 MILITARY LABS

Military facilities often handle materials that can be potentially hazardous, making biosecurity measures vital. Additionally, the military is frequently involved in high-risk activities such as responding to bioterrorism events, or research on defense against biological threats. Therefore, the biosecurity measures in these facilities are typically stringent:

- **Strict Access Controls**: Military facilities usually have stringent controls on who can access areas where biohazardous materials are present. This can include security clearance requirements, escorts for visitors, inventory of VBMs, and strict logging of all personnel entering and exiting these areas.
- **Extensive Training**: Personnel in military facilities usually undergo extensive training in handling biohazardous materials. This includes training in the proper use of PPE, decontamination procedures, and emergency response protocols.
- **Advanced Containment Facilities**: Military facilities often contain labs with high biosafety levels (BSL-3 or BSL-4), with multiple layers of physical containment to prevent accidental release of biohazards.
- **Detailed Protocols**: Due to the high-risk nature of the work, military facilities often have very detailed and specific protocols for handling biohazardous materials. These are regularly reviewed and updated, and compliance is strictly enforced.
- **Emergency Response Capabilities**: Military facilities often have advanced capabilities for responding to biosecurity incidents, including specialized response teams and equipment.
- **Oversight and Regulation**: Military biosecurity operations are subject to oversight by various regulatory bodies to ensure safety and compliance with biosecurity standards. This can involve regular inspections and audits.
- **Research and Development**: Military facilities often research new biosecurity measures and technologies, such as improved detection and decontamination methods, or countermeasures against biological threats.

While the specifics can vary depending on the nature of the military facility and its mission, these general principles form the backbone of biosecurity measures in such settings. The goal is not only to protect the personnel working within the facility and the public, but also to ensure the military's operational readiness in the face of biological threats.

10.4 GOVERNMENT LABS

Government labs often conduct a wide variety of research, including public health research, environmental studies, agricultural development, and even defense-related studies. Thus, biosecurity measures will depend on the nature of the work being conducted, but some general principles can be applied, including:

- **Institutional Biorisk Committee (IBC)**: Like universities and large corporations, government labs should have an IBC to review and oversee all work involving biohazardous materials. The IBC would be responsible for ensuring that all work complies with relevant regulations and guidelines.

- **Training**: All personnel working in government labs should receive appropriate biorisk management training, which should be refreshed regularly and whenever new equipment or procedures are introduced.
- **Facilities and Equipment**: Government labs conducting high-risk work should have the appropriate containment measures, facilities, and equipment. This should include certified BSCs for handling biohazardous materials, validated autoclaves for sterilizing equipment and waste, and PPE for laboratory personnel.
- **Risk Assessment**: Before any work involving potentially biohazardous materials is conducted, a thorough risk assessment should be performed. This should identify potential hazards, evaluate the likelihood and potential impact of exposure, and define appropriate risk management measures.
- **Emergency Response Planning**: Government labs should have detailed plans for responding to emergencies, such as spills, exposures, or equipment failures. This should include procedures for containing the incident, treating exposed personnel, notifying relevant authorities, and conducting a post-incident review to prevent future occurrences.
- **Public Communication**: Given their public nature, government labs have a responsibility to communicate openly and transparently about their work, and the measures they are taking to manage biosecurity. This could involve regular reports, public meetings, or a designated point of contact for public inquiries.

Implementing these biosecurity measures can help government labs to conduct their work safely and responsibly, while also building public trust in their activities.

10.5 CLINICAL STUDIES CENTERS

Clinical study centers, where new drugs, medical devices, and treatment procedures are tested in clinical trials, need biosecurity measures to protect the study participants, the healthcare providers, and the integrity of the research. Some key measures that are typically implemented include:

- **Training**: Staff at clinical study centers should receive comprehensive training in biorisk management. This includes understanding potential biohazards, the correct use of PPE, decontamination procedures, and how to respond to emergencies.
- **Informed Consent**: All participants in clinical trials should be fully informed of any potential biohazards and the measures taken to mitigate them. This is part of the informed consent process, which is crucial for ethical clinical trials.
- **Personal Protective Equipment (PPE)**: Clinical study centers should have adequate stocks of PPE, and all personnel should be trained in its correct use. The level of PPE required will depend on the nature of the clinical trial and the potential biohazards involved.
- **Risk Assessment**: Before any clinical trial is conducted, a thorough risk assessment should be conducted. This should consider any potential

biohazards, the likelihood of exposure, and the potential impact if exposure occurs.

- **Safe Handling and Disposal of Biohazardous Material**: Protocols should be implemented for the safe handling and disposal of any biohazardous material, such as body fluids, tissues, or used PPE.
- **Emergency Response Planning**: Clinical study centers should have plans for responding to biosecurity incidents, such as accidental or intentional biohazard exposure. This should include first aid procedures, notification of public health authorities, and measures to prevent further exposure.
- **Institutional Review Board (IRB)**: An IRB should review and approve all clinical trials before they begin. The IRB will consider potential biohazards as part of its assessment of the risks and benefits of the trial.

By implementing these and other biosecurity measures, clinical study centers can protect their staff, their participants, and the broader public while conducting crucial research into new treatments and interventions.

10.6 NON-PROFIT RESEARCH CENTERS

Non-profit research centers, which can encompass a broad range of research areas including environmental, agricultural, and medical research, must also prioritize biosecurity. The specifics of the biosecurity measures implemented will depend on the nature of the work being done, but some general principles apply:

- **Risk Assessment**: Before initiating any project involving potentially biohazardous materials, a comprehensive risk assessment should be conducted to identify potential hazards, evaluate the likelihood of exposure, and develop appropriate risk mitigation strategies.
- **Institutional Biorisk Committee (IBC)**: Forming an IBC is crucial for reviewing and overseeing all work involving biohazardous materials. The IBC ensures compliance with regulatory requirements and reviews research protocols for safety and potential impacts.
- **Training**: All staff and researchers should receive appropriate training in biorisk management practices. This training should cover general lab safety, specific practices for handling biohazardous materials, the correct use of PPE, and emergency response procedures.
- **Facilities and Equipment**: The availability and proper maintenance of safety equipment and facilities, such as biological safety cabinets, containment laboratories, and proper waste disposal systems, are crucial for mitigating biohazards.
- **Emergency Response Planning**: Non-profit research centers should have a well-detailed plan for responding to biosecurity incidents. This plan should outline steps for immediate containment, treatment of exposed personnel, and reporting to the relevant authorities.
- **Public Communication and Transparency**: Given their public-oriented missions, non-profit research centers have a responsibility to communicate

openly about their work and the measures they are taking to manage biosecurity. Regular reporting, public meetings, and having designated points of contact for public inquiries can contribute to transparency and trust-building.

These biosecurity measures can help non-profit research centers to conduct their work safely and responsibly, thus protecting their personnel, the public, and the environment.

10.7 MULTINATIONAL COLLABORATIVE CENTERS

Multinational collaborative centers, where researchers from different countries collaborate on various projects, face unique challenges in biosecurity. The differences in regulatory frameworks, resources, cultures, and language among the collaborating entities will impact the implementation of biosecurity measures. However, the following strategies can be applied to ensure effective biosecurity:

- **Harmonized Biosecurity Protocols**: These centers should strive to harmonize their biosecurity protocols to the highest standard among the participating entities. This includes protocols for handling, storing, and disposing of biohazardous materials, emergency response, and incident reporting.
- **Shared Responsibility**: Biosecurity should be considered a shared responsibility among all participating entities. There should be clear guidelines on the roles and responsibilities of each entity in managing biosecurity.
- **Training**: All personnel involved in the collaboration should receive appropriate training in biosafety and biosecurity. The training materials and guidelines should be available in the languages spoken by the participating entities.
- **Effective Communication**: There should be clear and effective communication channels among all entities involved in the collaboration. This includes communication about biosecurity protocols, incident reporting, and updates on regulatory changes in any of the countries involved.
- **Risk Assessment**: A thorough risk assessment should be conducted before any collaborative project begins. This should consider the specific biohazards associated with the project and the unique risks posed by the multinational nature of the collaboration.
- **Regulatory Compliance**: The collaboration must comply with the biosecurity regulations of all countries involved. This might require obtaining necessary permits, especially for transferring biohazardous materials across borders.
- **Trust and Transparency**: Given the multinational nature of the collaboration, trust and transparency are crucial. All entities should be open about their practices, and there should be transparency in decision-making processes.

Multinational collaborations can significantly advance scientific knowledge and innovation, but they require careful management to ensure biosecurity. By implementing these measures, multinational collaborative centers can safely conduct their essential work.

10.8 GARAGES, DIY LABS, iGEM COMPETITIONS, AND HIGH SCHOOLS

Biosecurity is often overlooked in nontraditional research environments such as garage labs, DIY biology workshops, iGEM competitions, and high schools, yet it is important to implement it in these settings. A one-size-fits-all approach to biosecurity is ineffective because of the different risks and requirements of each setting. Instead, tailored biosecurity measures that extend past mere protocols are important for establishing a persistent culture of biosecurity:

- **Garages as Research Spaces**: Garages, often seen as informal research spaces, are increasingly becoming venues where scientific activities take place. Lacking the strict biosecurity measures and oversight observed in formal labs, these spaces are particularly vulnerable to accidents or misuse of biological agents. Therefore, having basic biosecurity measures is important.
  - **Case Study**: A recent example of dangers about operating a "garage-like" lab space presented itself in the town of Reedley, California, where the discovery of an illegal medical lab has drawn significant attention from various levels of government. The lab was found to be producing and selling pregnancy and COVID tests online. Upon investigation, officials uncovered multiple code violations and a disturbing scene – approximately 1,000 mice, 200 of which were dead, and refrigerators storing infectious agents like chlamydia, *Escherichia coli*, human herpes, HIV, and SARS-CoV-2 strains without approval or oversight. The facility was promptly shut down in March 2023.[163]
  - **Biosecurity Lesson**: From a biosecurity perspective, the Reedley lab incident serves as a stark reminder of the risks posed by unauthorized and unregulated biological research facilities. These risks include the potential for accidental release of pathogens and the misuse of biological agents, which could lead to public health crises. The incident underlines the importance of robust oversight and regulation of biotechnological endeavors to ensure compliance with safety and ethical standards. It also highlights the need for transparency in funding and operations to prevent the diversion of resources to illicit or unsafe practices. Vigilance in monitoring and enforcement is crucial, and this case demonstrates the value of interagency cooperation in addressing biosecurity breaches. Prompt action in the face of such violations will help mitigate risk and uphold public trust in biosecurity measures.
- **DIY Biology Labs**: DIY biology labs pose similar challenges, as they frequently operate outside the jurisdiction of established academic institutions. Though fostering creativity and hands-on experience, inadvertent or intentional hazardous activities could take place there if not managed correctly.
  - **Case Study**: A recent example of DIY Biology involves a biohacker in California, who has gained notoriety for his public experiments, including injecting himself with CRISPR gene-editing technology during a live-streamed event. He operates a company that sells biohacking

supplies out of his garage, including comprehensive genetic engineering kits. However, his high-profile activities have attracted legal scrutiny, and he is now under investigation for allegedly practicing medicine without a license. An investigation by California health officials underlines the nascent legal challenges surrounding biohacking. The individual defends his actions by criticizing the FDA and government for limiting access to innovative treatments. Despite expressing regret over potentially inciting others to risky behaviors, he continues to sell his DIY CRISPR kits.[164]

- **Biosecurity Lesson**: This case exemplifies the biosecurity concerns inherent in the rapidly growing biohacking movement. The primary lesson here is the urgent need for regulatory frameworks that can keep pace with technological advancements and unconventional scientific practices outside of traditional laboratory settings. Biohacking poses unique risks, including the potential for self-harm, the uncontrolled release of genetically modified organisms, and the proliferation of DIY gene-editing capabilities without proper oversight. The incident underscores the importance of developing clear guidelines and ethical standards for biohacking activities to protect individuals and the public from the risks associated with unsupervised biomedical experimentation. Furthermore, it highlights the tension between innovation and regulation, illustrating the complexity of balancing the freedom to experiment with the need to ensure public safety. Ensuring that biohackers are well-informed about the possible consequences of their experiments, as well as that they operate within the bounds of the law, is crucial for the responsible development of biohacking and related fields.

- **iGEM Competitions**: In contrast, iGEM competitions, where teams of students work on synthetic biology projects, are generally hosted within academic institutions, which may already comply with existing biosecurity protocols. However, the global nature of these competitions brings into play regulatory and compliance variations across different countries. Standardizing biosecurity measures across all participating teams is, thus, crucial for mitigating risks and ensuring a level playing field.[165]
- **High Schools**: High schools represent an emerging frontier for biosecurity, with biology labs becoming more sophisticated and accessible. Considering that students and staff in these settings are generally less experienced in handling potentially hazardous biological material, it is imperative to have stringent biosecurity protocols. These should include collaboration with local colleges and universities that can provide oversight, training, guidelines on proper lab conduct, general laboratory safety, biosafety, and biosecurity.

10.9 CONCLUDING THOUGHTS

Implementing biosecurity measures using a graded approach across various environments, from academic institutions to industrial and non-traditional settings, is important to ensure public health and environmental integrity. The application of biosecurity measures in synthetic biology is a matter of responsible scientific research and a necessity for diverse applications. Addressing the critical need for biosecurity across various settings engaged in biological research and applications is essential. These settings range from high-level research laboratories to educational institutions, healthcare facilities, industrial environments, startups, and DIY labs, all underscoring the universal significance of biosecurity. Regardless of the scale or nature of the work, adequate protocols are crucial to prevent the accidental release of biological materials and to ensure the safety of all involved parties. The successful implementation of these measures is not just a matter of compliance but a fundamental aspect of ethical and responsible scientific practice. The proactive and informed approach called for across various settings is necessary and timely. Emphasizing continuous risk assessment and training reflects an understanding of biosecurity as an ongoing, evolving process. This approach is vital for adapting to new challenges and technologies, ensuring that biosecurity measures remain effective and responsive in a rapidly advancing field.

10.10 KEY TAKEAWAYS

- **Universal Importance of Biosecurity**: Emphasizes the necessity of biosecurity measures across all settings involved in biological research, especially those using synthetic biology.
- **Diverse Application**: Covers biosecurity implementation in various settings like research labs, educational institutions, healthcare facilities, and DIY labs.
- **Proactive Approach**: Advocates for continuous risk assessment, regular training, and awareness programs.
- **Cultural Shift**: Highlights the need for fostering a culture of biosecurity and responsibility.
- **Biosecurity and Compliance**: Stress the importance of implementing adequate biosecurity protocols to prevent theft or loss and ensure compliance with regulations.
- **Customized Measures**: Recognize the need for tailoring biosecurity measures to specific settings and activities.

10.11 THOUGHT-PROVOKING QUESTIONS

10.1 What would be the challenges of having larger universities oversee the biorisk management at nearby small colleges and high schools so that smaller institutions can safely train their students and simultaneously give students a mindset of safety practices?

10.2 How can we evaluate the efficacy of biosecurity measures implemented in an academic setting for assessing the probability of an accident?

10.3 How can we evaluate the efficacy of biosecurity measures implemented in an academic setting for assessing a prompt and efficient response in case of an incident?

10.4 Which parameters should be employed to evaluate the efficacy of biosecurity measures implemented in an academic setting for assessing the probability of an accident and ensuring prompt and efficient response in case of an incident?

10.5 What improvements made to training, regular auditing, updating measures based on new research, and better resource allocation have led to a reduction in the frequency and severity of biosecurity incidents?
10.6 What legal penalties should be enforced to justify the increasing costs and other resources needed for implementing up-to-date biosecurity measures?
10.7 What elements should be included in designing a biosecurity training program for students in an academic setting?
10.8 What arguments can be presented to the President of a research university for the role of students, faculty, and staff in implementing and maintaining biosecurity measures?
10.9 What role do various stakeholders play in implementing effective biosecurity measures in different settings? Who are the stakeholders?
10.10 What obstacles are encountered in the enforcement of biosecurity protocols, and which strategies could be employed to surmount them?
10.11 In what ways might university students with specialized training contribute during an emergency response?
10.12 At what juncture is an academic institution most susceptible to threats from within, and how can such vulnerabilities be mitigated?

11 Conducting a Biosecurity Risk Assessment

This chapter presents a systematic, step-by-step guide for conducting biosecurity risk assessments, which are a critical process for identifying and mitigating potential security risks associated with biological research, particularly in fields like synthetic biology. This chapter focuses on identifying vulnerabilities within laboratories and research facilities. It guides the reader in systematically evaluating security protocols, physical infrastructure, personnel reliability, and emergency response capabilities. The chapter stresses the need for a comprehensive approach encompassing all biosecurity aspects, including physical, personnel, and informational security. Following the assessment, the chapter discusses strategies for mitigating identified risks. This includes implementing appropriate security measures, developing response plans for potential biosecurity incidents, and regularly reviewing and updating risk management strategies. The chapter discusses integrating biosecurity risk assessments into managing biological research facilities. It highlights the need for ongoing training and awareness among staff and researchers, as well as the importance of fostering a culture of biosecurity and responsibility.

DOI: 10.1201/9781003423171-11

11.1 THE EIGHT BIOSECURITY PRIORITY AREAS

A robust Biorisk Management Program comprises two critical components: biosafety and biosecurity. Initially, biosecurity rested on five foundational elements: Physical Security, Personnel Management, Material Control and Accountability, Transport Security, and Information Security.[166] However, this framework has become antiquated, omitting crucial biosecurity dimensions. In response, Dutch experts have expanded this list to eight biosecurity pillars.[147] Subsequently, an international team led by Dutch experts refined these pillars, designating them as eight biosecurity priority areas.[157] These areas are essential for developing and administering a comprehensive Biosecurity Program within the broader context of an integrated and comprehensive Biorisk Management Program (see Figure 11.1). These eight biosecurity priority areas are:

1. **Management**: Management, in biosecurity, refers to the leadership and organization of Biosecurity Programs and measures within a particular institution or organization. It typically includes decision-making

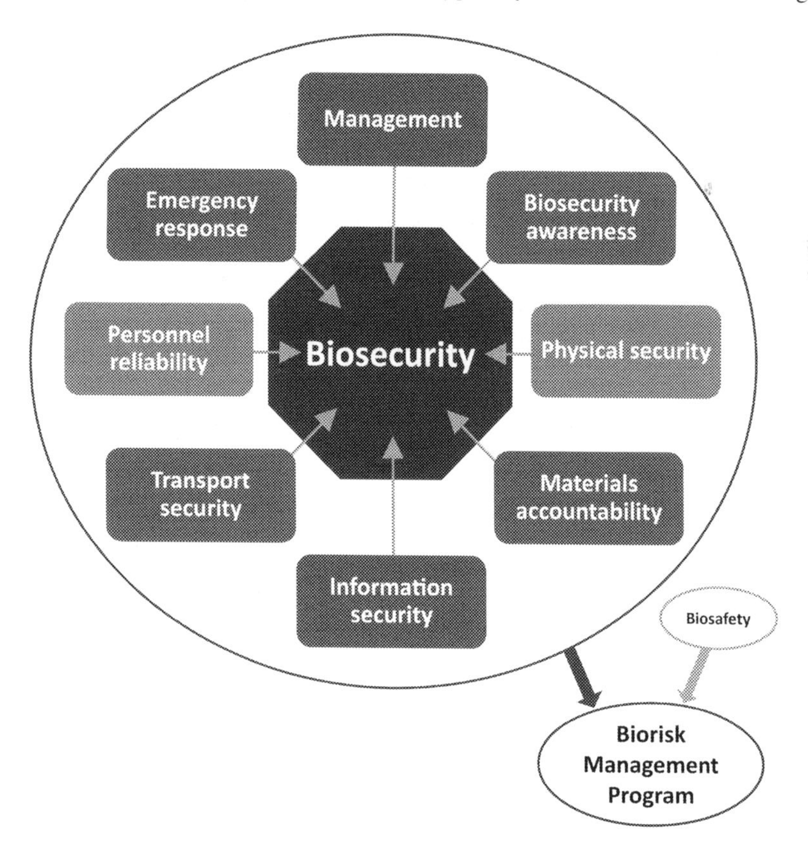

FIGURE 11.1 Eight biosecurity priority areas. Together, the eight biosecurity priority areas form a comprehensive biosecurity program. When integrated with a biosafety program, they contribute to a comprehensive biorisk management framework. The diagram emphasizes the relationship and importance of various security measures in maintaining biosecurity and managing biological risks.

processes, resource allocation, policy development, and oversight of biosecurity activities. The effectiveness of the management team is vital in establishing, implementing, and maintaining an effective Biosecurity Program. It is responsible for creating and enforcing biosecurity policies, as well as providing the necessary resources for the successful implementation of these policies.

2. **Biosecurity Awareness**: Biosecurity awareness encompasses the knowledge and understanding of biosecurity threats, measures, and protocols among personnel in an organization. It includes awareness of the risks associated with biological materials and the measures required to mitigate these risks. Regular training and educational programs are crucial for maintaining high levels of biosecurity awareness. Such programs keep personnel updated on potential threats, prevention measures, and procedures to follow in the event of a breach.
3. **Physical Security**: Physical security refers to the physical measures employed to protect facilities, personnel, and resources from unauthorized access and other potential threats. It includes locks, surveillance systems, access controls, and barriers. An effective physical security system is characterized by a combination of preventive and deterrent measures, which work together to protect facilities and their contents. These measures must be regularly assessed and updated to ensure they continue to offer robust protection.
4. **Materials Accountability**: Materials accountability involves keeping accurate and updated records of valuable biological materials (VBM). This includes tracking their acquisition, storage, transfer, and disposal. The key aspect of materials accountability is establishing a robust tracking system. This system allows for real-time monitoring of VBMs and facilitates quick response to any unexpected changes, thereby mitigating potential biosecurity risks.
5. **Information Security**: Information security refers to the protection of sensitive and confidential information related to the organization's biosecurity activities. This includes, but is not limited to, data about biological materials, personnel, research findings, and security protocols. Information security is achieved through access controls, encryption, and secure storage and transmission protocols. Regular audits and updates of information security measures are crucial to maintain effectiveness.
6. **Transport Security**: Transport security pertains to the measures taken to secure the transportation of biological materials to prevent unauthorized access, theft, or accidental release during transit. A comprehensive transport security plan encompasses pre-transport risk assessment, secure packaging, reliable transport means, and tracking mechanisms. It may also include contingency plans for incidents occurring during transit.
7. **Personnel Reliability**: Personnel reliability refers to the trustworthiness and dependability of personnel with access to biological materials and related information. This involves their adherence to biosecurity policies and procedures, as well as their mental and emotional stability. Ensuring personnel reliability involves processes such as background checks, training, and monitoring. It may also include measures to support the mental and emotional health and overall well-being of personnel.

8. **Emergency Response**: Emergency response involves the actions taken to address biosecurity breaches or biological incidents. This includes immediate containment measures, investigation, and recovery actions. An effective emergency response plan is characterized by clear roles and responsibilities, fast response times, and coordination between different stakeholders. It also includes after-action reviews to learn from incidents and improve future response efforts.

These priority areas must be underpinned by an overarching program management structure that supports, coordinates, and continually evaluates their effectiveness. Only through such a comprehensive and structured approach can an organization hope to mitigate biosecurity risks effectively. Table 11.1 shows lists of questions that can guide the reader when assessing a Biosecurity Program. These questions are based on the eight priority areas in a Biosecurity Program described above.

TABLE 11.1
Biosecurity Priority Area Assessment Questions

Biosecurity Priority Area	Questions
1. Management	1. Has the organization's senior management developed, authorized, and signed a policy on biorisk management? 2. Does the senior management of the organization actively participate in the biosecurity policy discussions and decisions? 3. Is the biosecurity policy periodically (annually) revised based on past experiences and risk assessments? 4. Has a specific budget or resources been allocated to managing biosecurity within the organization, and is the allocated budget sufficient to execute planned activities, including biosecurity training programs? 5. Does the organization have a system for monitoring unauthorized personnel conducting routine non-laboratory functions? 6. Does the organization have a system for conducting and reviewing biosecurity assessments, including risk assessments about to dual-use technology and other technological advancements? 7. Is compliance with procedures and conduct rules actively monitored within the organization, including having designated specific personnel to oversee the implementation of biosecurity measures? 8. Does the organization maintain an approved list of certified vendors and buyers for biological substances (VBM) including establishing a system to minimize the risk of procuring counterfeit items?
2. Biosecurity awareness	1. Does the organization conduct annual biosecurity awareness activities for all personnel, including those working in the laboratory, IT support, security support, and other supporting roles, including the foundations of biosecurity and their respective roles and responsibilities? 2. Does the organization have an ongoing training program for all personnel involved in implementing biosecurity measures, including how to respond to a biosecurity breach?

(*Continued*)

TABLE 11.1 (*Continued*)
Biosecurity Priority Area Assessment Questions

Biosecurity Priority Area	Questions
	3. Are awareness of dual-use technology and other technological advancements incorporated into the training and awareness programs? 4. Are personnel aware of the procedure for reporting any biosecurity breaches, and is the confidentiality of whistleblowers ensured? 5. Is the organization's Biosecurity Program understood and administered by competent personnel? 6. Do the personnel understand the response mechanism and their respective responsibilities in the event of a biosecurity breach? 7. Are personnel aware of their specific responsibilities about biosecurity and how these responsibilities are allocated? 8. Is there a friendly "see something say something" culture?
3. Physical security	1. Is there a policy for access controls that is enforced by management? 2. Are different areas within the facility subject to varying levels of security, where the access to secured areas, including storage locations of valuable biological materials (VBM) or select agents (SA), is protected through a combination of different security measures? 3. Are access controls monitored consistently across all secured areas, including secure storage for VBM or SA? 4. Does the facility employ an intrusion detection system to alert against unauthorized entry into the facility and storage areas for biological agents? 5. Are there surveillance cameras covering all external entrances to the laboratory building? Alternatively, is there 24-hour surveillance by security guards around the perimeter of the laboratory? 6. Are identification cards or badges required for all personnel and visitors to identify themselves within secure areas? Alternatively, are there guards at a unique entry point verifying individuals before entering the premises? 7. Is there an established procedure for escorting visitors within designated secure areas? 8. Do laboratory doors close automatically? 9. Is there a key control official overseeing the supervision and control of locks and keys for all buildings and entrances, and only distributing keys to authorized personnel?
4. Materials accountability	1. Does the organization have established policies for managing the inventory of valuable biological materials (VBM), including guidelines transferring VBMs and limits on the quantity of VBMs it holds? 2. Are there biosecurity procedures within the organization designed to prevent the intentional spread of biological agents, including having a designated individual responsible for registering and actively managing VBMs to ensure their control? 3. Does the organization consistently maintain and update its inventory records containing detailed information about the location of biological agents, including periodic reviews of the inventory of biological agents?

(*Continued*)

TABLE 11.1 (*Continued*)
Biosecurity Priority Area Assessment Questions

Biosecurity Priority Area	Questions
	4. Are there biohazard signs at laboratory entrances and storage areas that indicate the presence of biological agents without specifying which organisms are present? 5. Is the number of inventory storage locations minimized and sufficiently secured to ensure access by authorized personnel only? 6. Does the organization have established policies for the proper and secure disposal of VBMs, especially DURC VBMs? 7. Does the organization have a validated system for in-house disposal (through autoclave or incinerator)? 8. Does the organization have a system to trigger investigations in the event of unusual or suspicious activities?
5. Information security	1. Does the organization have an established and enforced policy related to information security, including using a classification system to identify sensitive information? 2. Does the organization maintain policies and procedures concerning individual access permissions for sensitive or confidential information, including designating authorized individuals responsible for overseeing information security? 3. Are all personnel well-versed and compliant with procedures related to accessing sensitive or confidential information? 4. Is sensitive or confidential information, including physical documents, secured in a safe location? 5. Are computers containing sensitive or confidential information password protected? 6. Do users have administrative access to their computers? 7. Are there computers that are not connected to the institutional network? Are individuals able to download potentially malicious programs? Are personnel susceptible to phishing scams? 8. Does the organization conduct regular exercises to identify vulnerabilities? Has the organization experienced hacking incidents in the past? 9. Has the organization installed appropriate security software on computers housing sensitive or confidential information? 10. Is there a system to back up sensitive or confidential information? 11. Are there established procedures for emergency response in case of a breach in information security? 12. Does the organization enforce administrative control measures for the exchange of sensitive information within the organization and between different institutions?
6. Transport security	1. Are individuals responsible for transporting valuable biological materials specifically trained in the procedures and requirements associated with transporting these materials? 2. Does the transport company chosen for this task adhere to relevant legal requirements? 3. Has the organization implemented a preselection process for transport companies intended to handle valuable biological materials?

(*Continued*)

TABLE 11.1 (*Continued*)
Biosecurity Priority Area Assessment Questions

Biosecurity Priority Area	Questions
	4. Does the organization enforce a chain of custody protocol? 5. Is there a tracking system for the transportation of biological samples? 6. Is a material transfer agreement established between the organization and the sender/recipient of valuable biological materials? 7. Does the organization ensure that the recipient institution possesses the necessary biorisk management capabilities to receive the samples safely? 8. Does the organization conduct risk assessments for each method of transportation utilized? 9. Does the organization have an emergency response plan when packages are lost during transportation?
7. Personnel reliability	1. Does the organization implement a personnel assessment protocol that includes conducting a thorough background screening for new hires, verification of credentials, skill evaluation, personal traits assessment, and risk-based background checks? 2. Are periodic background checks conducted on current employees? 3. Does the organization conduct mental health or psychological assessments for employees before employment or at regular intervals during their tenure? 4. Has the organization established guidelines and policies for visitors (such as students, contractors, clients, and temporary workers) regarding their security clearance to access the facility? 5. Does the organization maintain a current list of authorized personnel to access the facility and handle biological agents? 6. Does the organization provide guidelines and policies for employees to report or register unusual behavior of colleagues or visitors? 7. Does the organization have an evaluation system for cases where current employees are transferred to areas with a higher or lower risk profile? 8. Does the organization have a procedure for limiting or excluding personnel's access to the facility or biological agents based on risk assessment, applicable to temporary and permanent staff? 9. Does the organization have Standard Operating Procedures (SOPs) or guidelines to monitor the activities of employees working outside of regular business hours?
8. Emergency response	1. Has the organization established an emergency response plan that can effectively handle biological emergencies or breaches in biosecurity? 2. Are there clearly defined roles, responsibilities, and permissions outlined in the emergency response plan for handling and investigating biological incidents or emergencies? 3. Does the organization have a contingency plan to ensure day-to-day operations continue securely, especially in cases of emergencies? 4. Has the organization established communication protocols with relevant external parties to handle biological emergencies or breaches in biosecurity?

(*Continued*)

TABLE 11.1 (*Continued*)
Biosecurity Priority Area Assessment Questions

Biosecurity Priority Area	Questions
	5. Does the organization routinely conduct emergency drills or exercises incorporating biosecurity risks to confirm that personnel can respond appropriately to emergencies and other biosecurity situations?
	6. Has the organization set up protocols to rectify any breaches in biosecurity?
	7. Has the organization identified preventive measures or revised procedures to ensure that previous biosecurity breaches are not repeated?

This table provides a baseline for assessing the biosecurity management program at an institution. It includes guidelines and milestones to reach appropriate biosecurity levels. The stringency of implementation and frequency of measures should match the biosecurity risk level identified through a risk assessment.

11.2 BIOSECURITY RISK ASSESSMENT

Central to any Biosecurity Program is a thorough appraisal of the full spectrum of threats, from insider risks to external bad actors. A careful review of the biological materials, data, equipment, and other assets housed in a facility allows systematic analysis of their attractiveness for misuse by various actors. An accurate understanding of insider and outsider motivations, capabilities, and access opportunities is also vital. Characterizing information security gaps, theft likelihood, and potential consequences enables tailored safeguards against data and sample diversion.[157,167,168]

Following the steps of a comprehensive threat appraisal, a Biosecurity Program should be part of a holistic biorisk management program and unequivocally supported by the management or directorship of a laboratory or facility. It is crucial to remember that a robust Biosecurity Program is multifaceted, standing upon eight biosecurity priority areas. A good biosecurity risk assessment entails protocols and procedures for three distinct points in time, as illustrated in Figure 11.2:

1. **Pre-incident**: How to prevent an incident, implement mitigation procedures
2. **Incident**: What to do at the moment of an incident, emergency response
3. **Post-incident**: How to contain and mitigate the consequences of an incident, root-cause-analysis, and revised biosecurity risk assessment

11.3 THE FIVE-STEP PROCEDURE FOR CONDUCTING A BIOSECURITY RISK ASSESSMENT

A comprehensive biosecurity risk assessment involves a systematic, step-by-step procedure to evaluate threats, vulnerabilities, and consequences. This assessment begins with identifying the biological assets and materials to be protected, along with potential insider and external threats that may target them. Biological assets are divided into two categories: select agents (SA), which can be used for terrorism or with nefarious intent, and valuable biological materials (VBM), which may or may

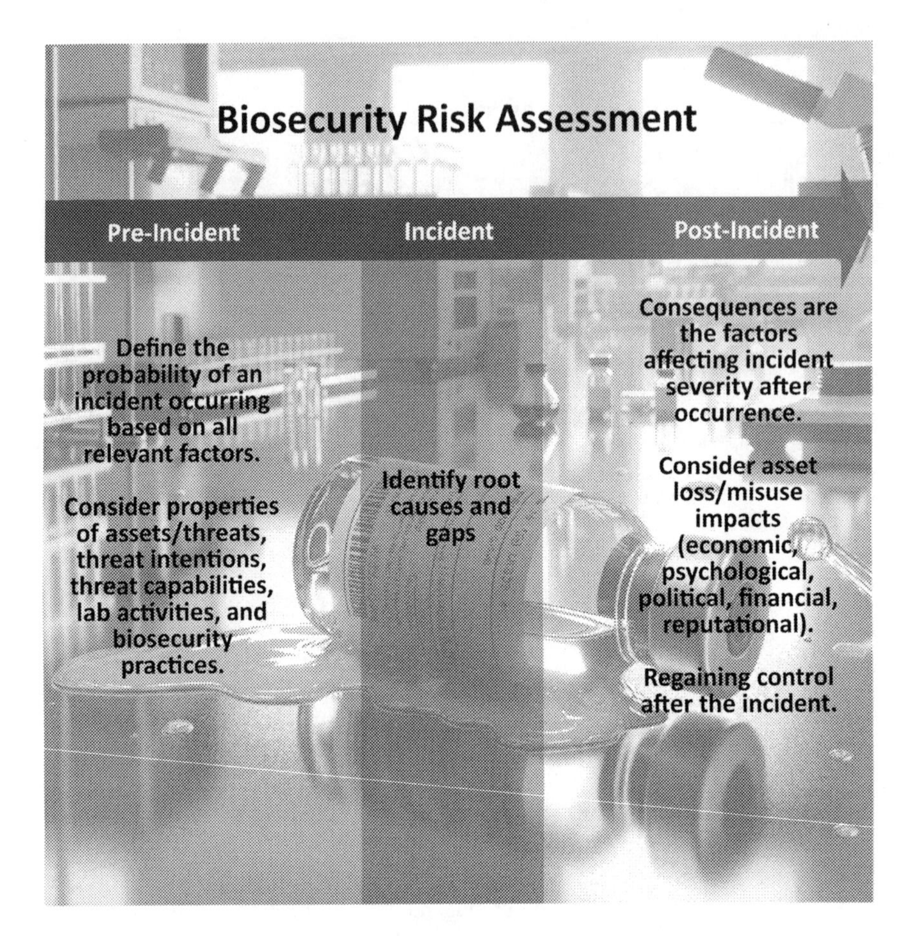

FIGURE 11.2 Biosecurity risk assessment of pre-, during, and post-incident. The pre-incident phase emphasizes determining the likelihood of an event. During the incident, the priority shifts to identifying root causes and gaps. The post-incident phase focuses on analyzing the consequences and understanding the contributing factors to the incident.

not be harmful but have intrinsic value monetarily, as IP, or as part of the research endeavor. A key consideration is characterizing the attractiveness of assets to different adversaries and the potential consequences of their theft or misuse. The risk assessment examines existing security measures and procedures to determine the likelihood of a successful security breach or theft based on current vulnerabilities. The overall biosecurity risk posed by various scenarios can be determined by evaluating the likelihood of a successful attack/theft and the severity of consequences together. This risk is often visualized via a matrix comparing likelihood categories (high/medium/low) against consequence categories (see Figure 11.4). The goal is to determine if risks are acceptable and, if not, which mitigation measures are needed to reduce vulnerabilities. Regular reviews and updating of the biosecurity risk assessments are essential as threats and science evolve.

TABLE 11.2
Basic Risk Assessment Process

Step 1	Define the circumstances	What work is occurring?
Step 2	Define the risks	What can go wrong?
Step 3	Characterize the risks	How likely is it to happen? What are the consequences? (see Figure 11.4)
Step 4	Determine if the risks are acceptable	Engage management and other key stakeholders
Step 5	Implement risk mitigation measures	Ensure all risks are acceptable post-implementation of mitigation measures

A risk assessment process involves the following basic steps.[96]

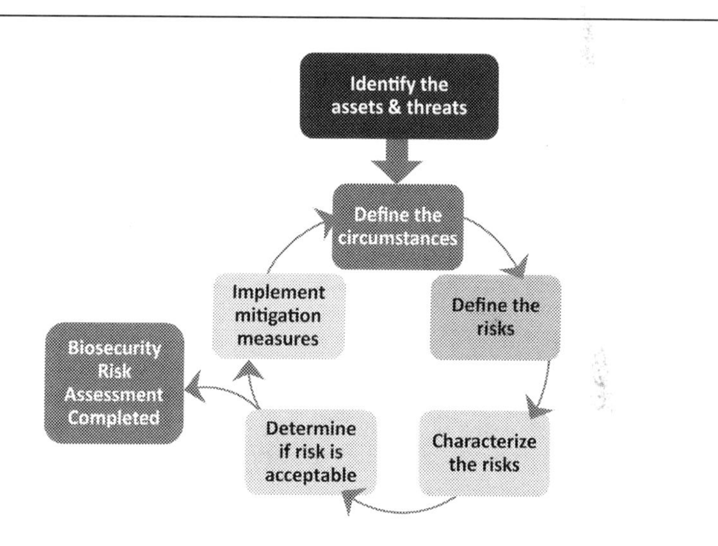

FIGURE 11.3 Flowchart describing the five steps in a biosecurity risk assessment process. The first step is to "Identify the assets and threats." Following this, the process flows through a cycle of five interconnected steps: (1) "Define the circumstances," considering the specific conditions that may affect the assets and threats. (2) "Define the risks," identifying the potential negative outcomes. (3) "Characterize the risks," evaluating the nature and magnitude of the risks. (4) "Determine if risk is acceptable," deciding whether the level of risk can be tolerated. If the answer is "yes" to the fourth step, then the cycle concludes. If the risk is deemed unacceptable, then it continues to step (5) "Implement mitigation measures," outlining actions to reduce or manage the risks. This represents the iterative and dynamic nature of risk assessment.

Here, I describe a step-by-step procedure, summarized in Table 11.2 and illustrated in Figure 11.3:

1. Define the circumstances (identify the biological materials and assets to be protected)
    a. Identify the assets (list the valuable biological materials, data, equipment, etc.)
    b. Identify the threats (determine potential insider and external adversaries)
    c. Define the facility and laboratory security environment (examine existing security measures)

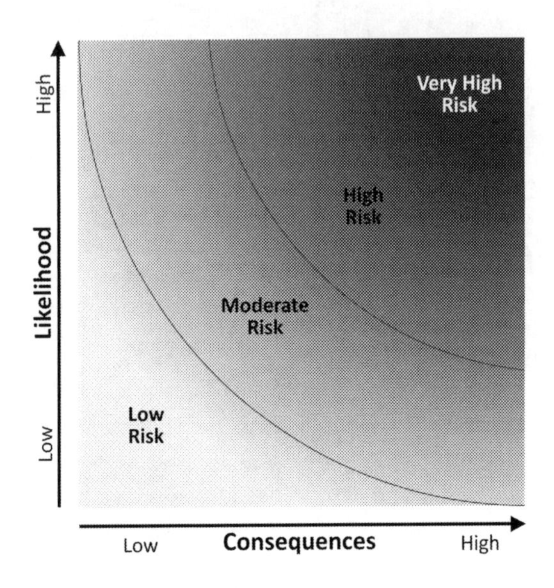

FIGURE 11.4 Risk Matrix for risk assessment. This is a graphical depiction of the "Likelihood vs. Consequences" of some event happening. At the bottom are the low risk, moderate risk, high risk, and very high risk. The graph is color-coded to represent different levels of risk. The light gray area closest to the origin where both likelihood and consequences are low represents "Low Risk." As the graph extends away from the origin, the risk level increases through shades of medium gray for "Moderate Risk" and "High Risk," eventually reaching dark gray for the "Very High Risk" area in the upper right corner, where both likelihood and consequences are high. The curved lines separating each risk level suggest a gradient, indicating that risk increases progressively as the likelihood of an event and its potential consequences increase. A similar graphical depiction can be obtained using the BioRAM program to conduct a biosecurity risk assessment,[105] or alternatively, it can be used manually to assess the risk qualitatively.

2. Define the risks (construct potential theft/misuse scenarios based on assets and threats)
3. Characterize the risks (evaluate likelihood and consequences, see Figure 11.4)
    a. Asset Assessment (determine attractiveness for theft/misuse)
    b. Adversary Assessment (evaluate adversary motivations and access)
    c. Facility Vulnerability Assessment (analyze avenues for compromise)
    d. Overall Risk Characterization (compare likelihood and consequences)
4. Determine if the risks are acceptable (consult with stakeholders on tolerability)
5. Implement risk mitigation measures, as needed (reduce vulnerabilities and enhance security)

Table 11.3 provides an overview of the biosecurity risk assessment process specifically, including specific details and what areas to assess. This is not an all-inclusive list and items should be added or removed depending on the circumstances and specifics of each assessment.

Example: See *Shigella*'s example (Section 11.4) for a detailed example of how to assess risk for a new project or research proposal in SynBio.

TABLE 11.3
Overview of the Biosecurity Risk Assessment Process

Steps	Specific Step Details	What to Assess
1. Define the circumstances	**1a.** Identify the assets	• Pathogens, toxins, equipment, intellectual property, lab animals, etc.
	1b. Identify the threats	• What are the adversarial types? (Insiders? Outsiders?) • What are their capabilities? • Motives? Means? • Opportunities?
	1c. Define the facility and lab security environment	• Review the biosecurity priority areas: 1. Management 2. Biosecurity awareness 3. Physical security 4. Materials accountability 5. Information security 6. Transport security 7. Personnel reliability 8. Emergency response
2. Define the risks	**2a.** Identify potential risks	• Construct scenarios for assets and adversaries
3. Characterize the risks	**3a.** Asset Assessment	• Assess the likelihood of assets being targeted • Consider the consequences of asset misuse or destruction
	3b. Adversary Assessment	• Evaluate the adversary's likelihood of success • Analyze the consequences of asset misuse or destruction
	3c. Facility Vulnerability Assessment	• Determine the likelihood of acquiring assets given vulnerabilities • Consider impact at laboratory, institutional, community, national, and global levels
4. Determine if the risks are acceptable	**4a.** Must be discussed with management and other stakeholders	• Include impacts on lab personnel, institution, and community
5. Implement mitigation measures, if needed	**5a.** Must be discussed with management and other stakeholders	• Engage the public and communicate within the institution and community for understanding and acceptance

The biosecurity risk assessment process involves defining risks, analyzing their likelihood and impact, evaluating if the risk is acceptable, and implementing risk mitigation. This table gives a general overview of the risk assessment process.[167]

11.4 HYPOTHETICAL SCENARIO – CONDUCTING A BIOSECURITY RISK ASSESSMENT WITH *SHIGELLA* SP.

Case Study: Hypothetical Development of a Synthetic *Shigella* Strain with Enhanced Pathogenicity for a Dual-Use Research of Concern (DURC) Experiment.[b]

Background on *Shigella*: In this hypothetical case, researchers aim to use synthetic biology techniques to create a synthetic *Shigella* strain with enhanced pathogenicity to understand bacterial virulence mechanisms better and inform the development of novel therapeutics. The scientists will then use a novel mouse model to study the new strain.[169] However, this experiment falls under the category of DURC due to the potential for misuse of the knowledge or techniques involved.

11.4.1 Risk Assessment Background Information: *Shigella* sp.

When gathering pathogen data and information, use research published in the scientific literature, resources available from national and international organizations, textbooks, and case studies. The case study below uses information from the Canadian Pathogen Safety Data Sheet (SDS) and is summarized in Table 11.4.[170]

Shigella is a Gram-negative bacterium that causes shigellosis, an infectious diarrheal disease.[170]

11.4.2 Biosecurity Risk Assessment for *Shigella* sp.

Biosecurity risk assessment is an essential component of synthetic biology research, particularly when working with potentially dangerous organisms like *Shigella* sp. Tools are available to help make the risk assessment process less subjective and more quantitative, such as using the freely available BiosecurityRAM program:[155]

- **Unintended Consequences**: There are two main types of unintended consequences to consider in this scenario. Within the lab, researchers could accidentally be exposed to the synthetic *Shigella* strain, which could lead to infection and disease. Outside the lab, accidental release of the strain into the community could cause an outbreak of disease that is difficult to control, especially if the strain is more virulent or antibiotic-resistant than naturally occurring *Shigella* strains.
- **Securing & Monitoring Materials**: Ensuring the secure storage and handling of the synthetic *Shigella* strain is crucial. This includes using appropriate containment facilities, implementing stringent access controls, and maintaining detailed records of when and by whom the strain is accessed. Regular audits can help to identify any discrepancies or security breaches promptly.
- **Emergency Protocols**: Laboratories must have robust emergency response plans. These plans should cover a range of scenarios, including accidental exposure of lab workers, unintended release into the environment, and intentional theft. Protocols might include immediate containment and

[b] This is a hypothetical scenario created for the purpose of illustrating how to do a biosecurity risk assessment. To the best of the author's knowledge at the time of publication, this is completely fictional.

TABLE 11.4
Agent-Specific Information for *Shigella* sp.

	Agent-Specific Information
Host range	Humans & higher primates
Infectious dose	10–200 organisms
Mode of transmission	Oral-fecal route
Incubation period	1–7 days
Communicability	Yes, through human feces, up to 4 weeks after infection
Zoonosis	None
Vectors	None, but survives on flies
Survivability outside host	• Months on dry surfaces • Days on contaminated vegetables • 2–28 days on metal utensils • 12 days on feces • 3 days in water • 2–24 days on flies
Inactivation	1 hour on autoclave, 70% ethanol, 1% sodium hypochlorite (bleach)
Drug susceptibility	Antibiotics, multidrug resistance has been resurfacing
Lab-acquired infections	Most common lab-acquired infection due to high virulence and low infectious dose
Toxins	Stx1 and Stx2 cause dysentery • Binding affinity, thermolability

This table summarizes the main agent-specific information required to perform a biosecurity risk assessment for a hypothetical synthetic biology experiment. As a guideline, when working with a new or modified organism, use the genus/species that is most genetically closely related to the organism to be created/mutated until the new organism is characterized.

decontamination procedures, medical treatment plans, notification procedures for relevant authorities, and investigation procedures to identify the cause and prevent recurrence.

- **Inter-Institutional Agreements**: Agreements with local government authorities and neighboring research institutions are a critical part of emergency preparedness. In the event of an emergency, a coordinated, rapid response is crucial. These agreements should clarify the roles and responsibilities of each party, establish communication protocols, and ensure access to the necessary resources for containment and treatment.

Given the interconnectedness of human, animal, and environmental health – a concept known as One Health – these agreements should also consider potential impacts on animal and environmental health and involve relevant veterinary and environmental agencies.

By considering these points, researchers can ensure a thorough biosecurity risk assessment, mitigating the potential dangers of handling and experimenting with synthetic *Shigella* strains.

Table 11.5 summarizes the biosecurity risk assessment in the hypothetical *Shigella* sp. scenario.

TABLE 11.5
Biosecurity Risk Assessment Process Summary for Hypothetical *Shigella* sp. Scenario

Steps	Specific Step Details	What to Assess
1. Define the circumstances	1a. Identify the assets	• Pathogen: *Shigella* sp. • Toxin: Shiga toxins 1 and 2 (Stx1 and Stx2) • Equipment: microscopes, sequencer, thermocyclers • Intellectual property: DNA sequencing data from the experiments, directed evolution analysis, lab experimental results, and protocols • Lab animals: a mouse model

	1b. Identify the threats	• What are the adversarial types?
		  • Insiders: Lab personnel, two lab members have been competing for projects and they both wanted to work on this one
		  • Outsiders: Competitors, lone actors, and terrorist groups
		• What are their capabilities?
		  • Insiders: The insiders have access to the lab facilities, equipment, strains, and lab animal facility
		  • Outsiders: The location is an "open campus," and the building is usually unlocked, although it has the capability of being locked
		• Motives?
		  • Internal: Personal gain of prestige, career advancement
		  • Outsiders: Media attention to highlight their worldview, disruption of work
		• Means? Taking out a tube with cells is easy; transporting mice would be harder to conceal
		• Opportunities? Both have opportunities to access the lab spaces as the campus is open and the building is often unlocked. With enough knowledge of people's schedules, they could access the spaces

	1c. Define the facility and lab security environment.	• **Physical security**:
		  • Asset security: Freezers can be locked, but are not typically locked
		  • Room security: The lab is locked and accessible by badge access
		  • Building security: There is after-hours guard security, but not specifically on this building door. Maintenance staff have access
		  • Perimeter security: None/minimal
		• **Personnel security/reliability**: Lab personnel is not formally screened prior to laboratory access, only interviewed by PIs to determine if they are competent to perform job duties, no background checks or formal ongoing behavioral assessments, and some informal behavioral assessments have been documented

(*Continued*)

TABLE 11.5 (*Continued*)
Biosecurity Risk Assessment Process Summary for Hypothetical *Shigella* sp. Scenario

Steps	Specific Step Details	What to Assess
		• **Material control and accountability**: Staff receive introductory inventory management training. There are no inventory systems; most groups do not keep inventory of materials; the ones that do have a spreadsheet on a computer in the lab that is not password protected. Inventory reconciliation is never conducted • **Information security**: Staff receive introductory training on information protection. They have admin controls over their computers. Computers are not always on the institutional network. They can download programs containing potential malware. They are vulnerable to phishing scams. The organization does not run routine exercises to assess vulnerabilities. They have not been hacked before. • **Transport**: Material transported across campus by lab staff on foot, usually in boxes or plastic containers. Shipments are done through the postal service of delivery companies. Packages could be potentially stolen or mistakenly delivered to the wrong location
2. Define the risks	**2a.** Identify potential risks.	• Construct potential scenarios considering various assets and adversarial types: • Risk of an insider person stealing valuable biological material for malicious purposes • Risk of an employee upset with a professor over losing a high-profile project, potential publication, and promotion. They intend to sabotage the project by contaminating strain stocks • Risk of an outsider stealing valuable biological material for personal use • Gain: A criminal intending to steal and sell a biological material or equipment which contains biological material for financial gain to an adversary organization • Risk of an outsider stealing equipment: A criminal intending to steal and sell a computer • Risk of an insider person stealing equipment: An employee stealing a computer for personal use • Risk of an outsider stealing an institution's intellectual property (in the form of information) or confidential information: An outside hacker intending to steal information and hold information hostage for ransom

(*Continued*)

TABLE 11.5 (*Continued*)
Biosecurity Risk Assessment Process Summary for Hypothetical *Shigella* sp. Scenario

Steps	Specific Step Details	What to Assess
		• Risk of an insider stealing an institution's intellectual property (in the form of information) or confidential information: A disgruntled employee desiring to sabotage the institution's reputation by leaking confidential information or deleting important files
3. Characterize the risks	**3a.** Asset assessment	• Define the likelihood of the asset being targeted: • HIGH for the disgruntled employee • MODERATE for other laboratory personnel (insiders) • LOW for outsiders • Consider the consequences of malicious use or destruction of the asset: • Malicious release of *Shigella* sp. – HIGH • Accidental exposure due to theft of equipment contaminated with *Shigella* sp. – MODERATE • Theft of Equipment – LOW • Loss of *Shigella* sp. details – MODERATE
	3b. Adversary assessment	• Define the likelihood of an adversary successfully targeting the asset: • Insiders: Not likely to consider stealing *Shigella* sp. for misuse toward the public, but more likely to steal the research details for personal gain • Outsiders: Not likely to consider stealing *Shigella* sp. for misuse by the public, but likely to steal IP and other valuable items. May cause accidental release due to theft of contaminated equipment • Consider the consequences of an adversary maliciously using or destroying the asset: • Insiders: HIGH, possible harm if others become infected, loss of IP • Outsiders: HIGH, possible harm if others become infected, loss of IP
	3c. Facility vulnerability assessment	• What is the likelihood of successful acquisition of the asset based upon the facility's vulnerabilities: • Physical Security: HIGH due to freezers not consistently locked, open campus and open building, access to all laboratory staff and maintenance crews, if necessary • Personnel Reliability: HIGH due to lack of background checks • Material Control and Accountability: MODERATE due to minimal training and no inventory system • Information Security: MODERATE due to minimal training

(*Continued*)

TABLE 11.5 (*Continued*)
Biosecurity Risk Assessment Process Summary for Hypothetical *Shigella* sp. Scenario

Steps	Specific Step Details	What to Assess

		• Transport Security: Within campus, MODERATE due to the informal nature of transport; outside of campus, LOW due to strict adherence to all rules and regulations
		• What is the likelihood that the asset being targeted by the threat, misused or stolen by the adversary, and successfully acquired by them?
		  • What is the likelihood of targeting the asset by the threat? MODERATE TO HIGH
		  • What is the likelihood of the adversary interest or desire to attempt to acquire the asset? LOW
		  • What is the likelihood of successful acquisition of these assets? MODERATE TO HIGH
		  • What is the overall likelihood of this occurring? MODERATE to HIGH
		• Combine the consequences of misuse of the asset
		• What is the consequence of the hazard causing harm? MODERATE to HIGH
		• What are institution consequences? MODERATE
		• The overall risks are:
		  • Risk of an insider stealing valuable biological material for malicious use – HIGH
		  • Risk of an outsider stealing valuable biological material for personal gain – LOW
		  • Risk of an outsider stealing equipment – LOW
		  • Risk of an insider stealing equipment – MODERATE
		  • Risk of an outsider stealing an institution's intellectual property (in the form of information) or confidential information – MODERATE
		  • Risk of an insider stealing institution intellectual property (in the form of information) or confidential information – HIGH

4. Determine if the risks are acceptable	**4a.** Must be discussed with management and other stakeholders	• Must consider lab personnel, institutional personnel, and the community: • Based on biosecurity practices when working with a dangerous biological agent (and possible DURC), these risks would be unacceptable, and additional mitigation measures would be required

(*Continued*)

TABLE 11.5 (*Continued*)
Biosecurity Risk Assessment Process Summary for Hypothetical *Shigella* sp. Scenario

Steps	Specific Step Details	What to Assess
5. Implement mitigation measures, if needed	**5a.** Must be discussed with management and other stakeholders	• Mitigation measures should include additional access controls, background checks for all laboratory employees with access to the agent, and implementation of an inventory management system including regular reconciliation of the material and strengthening of cyberbiosecurity

11.4.3 Biosecurity Concerns

Biosecurity concerns are paramount when dealing with synthetic biology, particularly when engineering organisms like *Shigella* with enhanced pathogenicity. This pursuit necessitates an intricate balance between facilitating scientific progress and ensuring societal safety. The following points outline the key elements of this discussion:

- **Human Factors**: Understanding the human elements involved in the research, production, and handling of a synthetic *Shigella* strain is critical. This includes proper training to ensure researchers have the necessary skills to handle the pathogen safely. Background checks and regular "fit for duty" assessments should be implemented to screen personnel for any factors that might compromise safety, such as health conditions or unmanaged stress.
- **Situational Awareness**: Cultivating a deep understanding of the biological properties of the synthetic *Shigella* strain and the potential implication of its misuse is crucial. Researchers and relevant personnel should be aware of potential routes of accidental release or theft and have contingency plans.
- **Insider Threat**: Given the potential for misuse, researchers and organizations need to be vigilant against the insider threat, which involves a person within the organization intentionally causing harm. Strong security measures, strict access controls, and continuous monitoring can help mitigate this risk.
- **Promoting a Healthy Safety Culture**: Organizations should actively promote a strong culture of safety and security. This includes regularly reinforcing safety protocols, encouraging open communication about safety concerns, and fostering a respectful, supportive environment where staff feel comfortable raising issues.
- **Development of a Synthetic *Shigella* Strain**: Developing a more virulent or antibiotic-resistant *Shigella* strain has severe public health implications. *Shigella* is already a significant global health concern, and enhancement of its pathogenic traits could lead to outbreaks that are difficult to control and treat. There is also a risk that such a strain could be misused for bioterrorism, with devastating consequences.

Considering these concerns, the biosecurity measures surrounding the development and handling of a synthetic *Shigella* strain must be stringent and thorough. These should include robust physical security measures, comprehensive personnel vetting and training procedures, and strict data and information security controls. In addition, there should be a strong regulatory framework with clear oversight and accountability mechanisms to ensure the responsible conduct of this type of research.

11.4.4 Measures Taken to Address Biosecurity Concerns

- **Review and Oversight**: The research proposal would be subject to review and approval by IBCs, relevant regulatory authorities, and DURC oversight committees. These committees should ensure that the proposed research meets the highest safety and ethical standards and that the potential benefits outweigh the risks.
- **Laboratory Containment**: The research would be conducted in a Biosafety Level 3 (BSL-3) laboratory, which is designed for handling highly pathogenic organisms. Containment measures would include strict access control, specialized ventilation systems, and rigorous safety protocols to prevent accidental exposure or release of the synthetic *Shigella* strain.
- **Risk Mitigation Strategies**: The researchers would need to develop and implement risk mitigation strategies, such as engineering the synthetic *Shigella* strain with built-in safety features (e.g., auxotrophies or inducible kill switches) to limit its ability to survive and spread outside the laboratory environment. Also, keep and maintain an updated inventory of stocks.
- **Limited Dissemination of Information**: To minimize the risk of misuse, specific details of the synthetic *Shigella* strain's genetic sequence, the techniques used to create it, and the resulting experimental data could be withheld or shared only with a limited number of researchers and public health officials with a legitimate need for access.
- **Collaboration**: Researchers should work closely with international partners, such as the World Health Organization (WHO), to share information and develop guidelines for responsibly conducting DURC experiments.
- **Promotion of a Culture of Responsibility**: The scientific community would be crucial in promoting a culture of responsibility, ensuring that researchers understand the potential risks associated with DURC and follow established codes of conduct and ethical guidelines.

11.5 CONCLUDING THOUGHTS

A thorough and systematic approach to conducting biosecurity risk assessments is vital for maintaining biosecurity in synthetic biology. This approach, as detailed, involves a meticulous process of identifying vulnerabilities within laboratories and research facilities. It guides through a comprehensive evaluation of security protocols, physical infrastructure, personnel reliability, and incident response capabilities.

Such a methodical approach is essential for preemptively identifying and mitigating potential threats, underscoring the dynamic nature of biosecurity risk assessment. This chapter provided a practical, step-by-step process for conducting a biosecurity risk assessment. It included a fictional DURC scenario to allow the readers to put into practice the guidance provided. The chapter's emphasis on a comprehensive approach, which includes physical, personnel, and informational security aspects, stresses the significance of an all-encompassing biosecurity strategy. Fostering a culture of biosecurity and responsibility among staff and researchers is highlighted as a critical factor. This cultural aspect ensures that biosecurity measures are enforced and ingrained in the daily practices of those involved in synthetic biology. Moreover, the focus on regular reviews and integrating risk assessment into managing biological research facilities reflects an understanding of biosecurity risks' dynamic and evolving nature. This holistic approach to biosecurity risk assessment and management is crucial for safely advancing synthetic biology and safeguarding research endeavors, public health, and environmental integrity.

11.6 KEY TAKEAWAYS

- **Systematic Risk Assessment**: Outlines a step-by-step guide for conducting comprehensive biosecurity risk assessments.
- **Identifying Vulnerabilities**: Focuses on identifying vulnerabilities within laboratories and research facilities.
- **Comprehensive Approach**: Emphasizes a holistic approach encompassing physical, personnel, and informational security.
- **Regular Review and Update**: Advocates for regularly reviewing and updating risk management strategies.
- **Cultural Importance**: Highlights the need for fostering a culture of biosecurity and responsibility among staff and researchers.
- **Integrating Assessments**: Stresses integrating biosecurity risk assessments into the management of biological research facilities.

11.7 THOUGHT-PROVOKING QUESTIONS

11.1 What policies and measures would you include in future biorisk management frameworks to promote transparency and open communication? Transparency and open communication might facilitate quicker responses to incidents, encourage trust among the public, and contribute to collective improvement in safety practices.

11.2 What cross-disciplinary contributions and steps can be taken in synthetic biology to prepare for and mitigate future pandemics?

11.3 How can biorisk management professional gauge if their plans and mitigation strategies are adequate via computer or virtual reality simulated outbreaks and other incidents?

11.4 What data would be needed to build an effective model to simulate synthetic biology accidents, assess their risk level to escalate into an outbreak, and develop mitigation measures?

11.5 What challenges have been consistently encountered in accident case studies? What biorisk management strategies can be developed to prevent an accident? What measures can be taken to mitigate the consequences of future accidents rapidly?
11.6 How can AI contribute to developing advanced tools for creating more precise and quantitative risk assessment models?
11.7 How could existing biosecurity risk assessment frameworks be enhanced to keep up with the swiftly developing field of synthetic biology? Specifically, what advancements would you suggest in areas such as quantitative models, including human factors, and dual-use concerns?
11.8 What are the critical considerations in doing a risk-benefit analysis to assess the potential biosecurity risks of an experiment involving an enhanced pathogen?
11.9 What components would you suggest for an oversight system that encourages responsible practices in dual-use synthetic biology research, promoting transparency and accountability while supporting valid research?
11.10 What novel strategies would you propose to enhance the biosecurity culture and foster responsibility among synthetic biology researchers?
11.11 What innovative methods would you recommend for the synthetic biology community to engage the public more effectively on dual-use issues?

12 Promoting Responsible Conduct in Biomedical Synthetic Biology

A thorough exploration of ethical and responsible conduct is required in biomedical synthetic biology. This chapter emphasizes the critical importance of maintaining high ethical standards and responsible practices in a field with significant potential for groundbreaking medical advancements and complex ethical dilemmas. The chapter delves into the ethical considerations unique to this field, such as the manipulation of genetic materials, the creation of novel organisms, and the implications of these technologies on human health and the environment. The chapter highlights the importance of considering the long-term societal impacts of synthetic biology research and the need for responsible stewardship of these powerful technologies. It discusses the role of education in promoting responsible conduct. This approach aims to build a solid ethical awareness and responsibility foundation from the early stages of scientific training. The chapter underscores the necessity of a concerted effort by the scientific community, regulatory bodies, and society at large to ensure that the advancements in biomedical synthetic biology are developed and

DOI: 10.1201/9781003423171-12

applied in a manner that is ethically sound, socially responsible, and beneficial to humanity. This commitment to responsible conduct is vital for maintaining public trust and fostering sustainable progress in this dynamic and impactful field.

12.1 EDUCATION AND TRAINING FOR SCIENTISTS AND RESEARCHERS

Education and training form a critical part of responsible practice in synthetic biology. In addition to technical knowledge and lab skills, scientists and researchers need to understand the ethical, social, and legal implications of their work. Key areas for education and training include:

- **Technical Skills**: This includes knowledge of molecular biology, genetic engineering techniques, bioinformatics, and other skills specific to synthetic biology. Hands-on lab training is crucial for developing these skills.
- **Biorisk Management**: Training in biorisk management practices is essential for any researcher working with biological materials. This includes understanding how to conduct risk assessments, use personal protective equipment, manage biohazardous waste, and respond to emergencies. Biosecurity training, which focuses on preventing misuse of biological materials, is also important.
- **Ethics**: Ethical training helps researchers understand the societal implications of their work and make decisions that respect the rights and welfare of others. Topics might include the responsible use of animals or human subjects in research, the ethics of genetic modification, and considerations for the fair distribution of benefits from synthetic biology.
- **Regulatory Knowledge**: Researchers need to understand the regulatory landscape for synthetic biology, including any requirements for risk assessment, reporting, or approval before certain activities can be undertaken.
- **Public Communication**: Skills in public communication can help researchers explain their work to a non-scientific audience, engage in dialogue about the potential benefits and risks, and build public trust in synthetic biology.
- **Dual-Use Awareness**: Training on dual-use issues can help researchers understand the potential misuse of their work and take steps to mitigate this risk.
- **Interdisciplinary Skills**: Synthetic biology is a highly interdisciplinary field, drawing on biology, engineering, computer science, and other disciplines. Training that breaks down disciplinary boundaries can be very beneficial.
- **Professional Development**: Besides technical skills, researchers can benefit from training in areas like project management, teamwork, leadership, and other "soft skills."

These topics should be incorporated into formal education programs for synthetic biology. Continuing education and professional development opportunities can help researchers keep up with new techniques, ethical debates, and regulatory changes in this rapidly evolving field.

12.2 WEAKNESS IN FORMAL EDUCATION AND TRAINING PROGRAMS REGARDING BIORISK MANAGEMENT

The effectiveness of biosafety and biosecurity measures depends on the individuals who implement and manage them. However, several critical gaps exist in the educational and professional landscape of biosafety and biosecurity that could jeopardize our ability to mitigate biological risks effectively.[171] This section explores three primary areas of concern: the limited availability and global inconsistency of formal education and training programs in biosafety and biosecurity, the worldwide shortage of biosafety professionals, particularly intensified by the COVID-19 pandemic, and the need for increased government funding in biorisk management research to advance these fields scientifically.

- **Formal Education and Training in Biorisk Management**: Despite the critical role professionals play in this field, only a select few universities offer structured programs in biorisk management. A few examples are institutions such as Johns Hopkins University[172] and George Mason University[173] in the United States, as well as Vietnam One Health University Network.[174] However, significant challenges persist. First, a standardization gap exists globally, leading to inconsistencies in competencies among biorisk management professionals across regions. Second, the salary packages offered often do not match the specialized technical expertise these professionals bring to their organizations, nor the critical role they play in preserving public health and the environment. Additionally, the significance of their work is often overlooked within their institutions. To bring about meaningful improvements in biorisk management protocols, it is crucial to grant these professionals a higher level of decision-making authority and better compensation.
- **Global Deficit of Biorisk Management Professionals**: The world currently faces a significant shortage of professionals in biorisk management, a problem aggravated by the COVID-19 pandemic. The capacity to manage and mitigate the risks of infectious diseases hinges on a strong, well-trained, and well-staffed biorisk management workforce. A shortfall in such professionals, therefore, presents a substantial risk to public health security, especially in developing regions where biorisk management infrastructure may be less developed.
- **Need for Government Investment in Biorisk Management Research**: Given the essential role of biorisk management in public health, academia, and industry, as well as the significant shortage of professionals in this field, governments worldwide need to invest more in biorisk management research. Such investments can help to advance the scientific understanding of biosafety and biosecurity, thereby enhancing credibility and respect

for this field among other scientific disciplines. Additionally, greater investment in biosafety and biosecurity can encourage more students to pursue careers in this field, helping to address the current shortage of professionals.

The U.S. government has expressed the country's vital need to focus on the biosafety and biosecurity of biological research and engineering. The CHIPS and Science Act calls for the White House Office of Science and Technology Policy to support "research and other activities related to the safety and security implications of engineering biology" and for the office's interagency committee to develop a strategic plan for "applied biorisk management" and to evaluate "existing biosecurity governance policies, guidance, and directives for the purposes of creating an adaptable, evidence-based framework to respond to emerging biosecurity challenges created by advances in engineering biology."[175]

Overall, addressing these weaknesses in formal education and training programs regarding biorisk management is crucial to meet the international need for more formalized training.[176] Doing so will help ensure a safer and more secure future and enable us to respond more effectively to emerging biological threats.

12.3 COLLABORATION AND INFORMATION SHARING

Collaboration and information sharing are vital for promoting biosecurity in synthetic biology. They ensure that lessons learned in one setting can be applied in another, and they help drive both technical advances and the development of best practices. Ways in which these principles apply include:

- **Sharing of Biosecurity Practices and Incident Information**: Labs and organizations that work with synthetic biology can learn significantly from each other's experiences. Sharing information about biorisk management practices, incidents, and lessons learned can help to improve safety across the field. This could be facilitated through online forums, workshops, or formal reporting systems.
- **Collaborative Development of Guidelines and Standards**: By working together, researchers and stakeholders can develop shared guidelines, standards, and best practices for synthetic biology. This can help to ensure consistency and promote a high level of safety across different labs and projects.
- **Joint Training and Education Initiatives**: Collaborative training programs can help to spread knowledge and skills related to synthetic biology safety. For example, universities, research institutions, and industry might partner to provide training courses, webinars, or workshops.
- **Shared Resources**: Sharing resources, such as safety protocols, training materials, or tools for risk assessment, can help to support biosecurity in labs that might not have the resources to develop these from scratch.
- **Open Science and Open Data**: Openness in science – including the sharing of data, methods, and results – can help to promote biosecurity by allowing others to validate and build upon existing work. However, it is important to

balance openness with the need to protect sensitive information that could potentially be misused.

- **International Collaboration**: Given the global nature of synthetic biology, international collaboration is crucial. This can help to share lessons learned from different countries, coordinate responses to global challenges, and build capacity in regions that are less experienced in synthetic biology.

By fostering a culture of collaboration and information sharing, the synthetic biology community can work together to continually improve safety, mitigate risks, and harness the benefits of this powerful technology.

12.4 PROFESSIONAL CODES OF CONDUCT AND ETHICAL STANDARDS

Professional codes of conduct and ethical standards provide important guidelines for individuals working in synthetic biology, ensuring that their work adheres to the highest standards of integrity, safety, and respect for life and the environment. Key elements in these standards often include:

- **Respect for Life and the Environment**: At its core, synthetic biology involves manipulating life at a fundamental level. Therefore, researchers need to have a deep respect for life and the environment, taking care to minimize harm and maximize benefits.
- **Biosafety and Biosecurity**: Professionals are expected to adhere to the highest standards of biosafety and biosecurity, taking all necessary precautions to prevent accidental release or harmful use of synthetic organisms. They should also stay informed about current best practices and regulations, as well as regularly review and update safety protocols.
- **Research Integrity**: Researchers should uphold the principles of honesty, accuracy, and transparency in all aspects of their work. This includes acknowledging the work of others, disclosing potential conflicts of interest, and correcting errors promptly when they are discovered.
- **Responsible Innovation**: Innovation in synthetic biology should be committed to the common good. This means considering the potential societal and environmental impacts of new technologies and engaging in dialogue with stakeholders to ensure that benefits are widely shared.
- **Openness and Collaboration**: Sharing of knowledge, data, and resources can help to advance the field and promote safety. However, openness must be balanced with protecting sensitive information and intellectual property rights.
- **Education and Mentorship**: Senior professionals have a responsibility to educate and mentor the next generation of synthetic biologists, instilling in them a solid commitment to ethical and professional conduct.

- **Public Engagement**: Professionals should strive to communicate their work to the public in a transparent and accessible way, and to listen to and respect public views.
- **Oversight and Accountability**: Professionals should cooperate with oversight bodies, report any biosafety, biosecurity, or ethical issues promptly, and take responsibility for the impacts of their work.

Ethical and professional standards provide a crucial framework for responsible practice in synthetic biology, helping to build public trust in the field and ensure that its benefits are realized safely and ethically.

12.5 PUBLIC ENGAGEMENT AND STAKEHOLDER INVOLVEMENT

Public engagement and stakeholder involvement are crucial components of biosecurity in synthetic biology. Ensuring that a wide range of perspectives is considered can help to identify potential risks, devise appropriate mitigation strategies, and build societal trust in the field. Ways in which this can be achieved include:

- **Public Consultation**: Public consultations can gather a range of views on proposed synthetic biology projects or policies. This can be achieved through surveys, public meetings, focus groups, or online platforms.
- **Stakeholder Engagement**: Besides the public, it is important to engage with specific stakeholders who might be affected by synthetic biology projects. This can include farmers, healthcare providers, environmental groups, industry, Indigenous communities, and others.
- **Inclusion of Diverse Perspectives**: Efforts should include diverse perspectives, including those of people from different geographical, cultural, and socioeconomic backgrounds. This can help to ensure that the benefits and risks of synthetic biology are distributed fairly.
- **Public Education**: Public understanding of synthetic biology can be enhanced through education initiatives, such as school programs, museum exhibits, science festivals, and media outreach. This can help to enable more informed and meaningful participation in decision-making.
- **Two-Way Dialogue**: Public engagement should be a two-way dialogue, not just a one-way information communication. This means not only informing the public about synthetic biology, but also listening to their views, concerns, and suggestions.
- **Transparency**: Being open and transparent about the aims, methods, and results of synthetic biology research can help to build public trust. This includes reporting on safety measures, incident responses, and potential risks.
- **Participatory Governance**: In some cases, members of the public or stakeholder groups might be involved directly in decision-making processes, such as through citizen juries, consensus conferences, or participatory technology assessment.

By engaging with the public and stakeholders in a meaningful and respectful way, the synthetic biology community can ensure that safety measures are responsive to societal needs and values and that the development and application of this technology are guided by the principles of social justice and democratic decision-making.

12.6 CONCLUDING THOUGHTS

In synthetic biology, particularly in its biomedical applications, intertwining ethical dilemmas with the need for responsible stewardship is important. This necessitates a comprehensive discussion that aligns scientific progress with ethical considerations. A key focus should be on education and training to instill ethical awareness from the early stages, preparing future scientists to navigate the moral complexities they may encounter. The imperative of maintaining high ethical standards and responsible practices is especially pronounced in a field already making groundbreaking medical advancements. Unique ethical considerations for experiments, such as manipulating genetic materials and creating novel organisms, must be thoroughly examined, particularly regarding their implications and unintended consequences for human health and the environment. It is necessary to consider long-term societal impacts and the need for responsible stewardship of these powerful technologies. This chapter makes an emphatic call for a concerted effort by the scientific community, regulatory bodies, biorisk management professionals, and society to ensure that advancements in biomedical synthetic biology are developed and applied in a manner that is ethically sound, socially responsible, and beneficial to humanity. This commitment to responsible conduct is essential for maintaining public trust and fostering sustainable progress in this dynamic and impactful field. A collaborative approach among various stakeholders reflects the multifaceted nature of the ethical challenges in synthetic biology, underscoring the necessity of diverse perspectives and inclusive dialogue in shaping the field's ethical landscape.

12.7 KEY TAKEAWAYS

- **Ethical and Responsible Conduct**: Emphasizes the importance of high ethical standards and responsible practices in biomedical synthetic biology.
- **Unique Ethical Considerations**: Discuss ethical issues specific to synthetic biology, including genetic manipulation and the creation of novel organisms.
- **Long-Term Societal Impacts**: Highlight the importance of considering the long-term impacts of synthetic biology research on society and the environment.
- **Role of Education**: Stresses the necessity of education in promoting ethical awareness and responsibility in synthetic biology.
- **Collaborative Effort for Ethical Advancements**: Calls for collaboration among the scientific community, regulatory bodies, and society to develop and apply synthetic biology advancements ethically and responsibly.
- **Building Public Trust**: Recognizes the need for responsible conduct to maintain public trust and sustainable progress in the field.

12.8 THOUGHT-PROVOKING QUESTIONS

12.1 What changes are needed to improve formal education and training programs in biorisk management so that officers/professionals have a broad range of competencies and increased decision-making authority?

12.2 What are the main challenges to improving the effectiveness of collaboration and information sharing in promoting biosecurity in synthetic biology domestically and internationally?

12.3 What areas should be prioritized when developing a proposal for a government strategy to invest in biosafety and biosecurity research, as well as to enhance the capacity and effectiveness of biorisk management measures?

12.4 What key elements should be included in a new "Codes of Conduct and Ethical Standards Guide" for biorisk management professionals? For Synthetic Biology Professionals?

12.5 At what stage in their educational journey should students be introduced to the concepts and management of biorisks?

13 Ensuring a Safe Future in the Age of Synthetic Biology

As extensively discussed in this book, addressing the biosecurity concerns within synthetic biology necessitates a comprehensive and collaborative approach. This approach must engage diverse stakeholders, including scientists, policymakers, risk management experts, and various other relevant parties. Achieving an equilibrium between the pursuit of scientific knowledge and this field's responsible advancement is imperative. Such balance is indispensable for harnessing the full potential of synthetic biology while mitigating the associated risks. We must approach these challenges with a forward-thinking mindset, mindful of the overarching principles of One Health. In this interconnected world, the consequences of our actions reverberate throughout the global ecosystem, affecting all living beings and the environment alike. As responsible custodians of our knowledge, natural resources, ethical application of knowledge, and technology, we are accountable for contributing to a more just and equitable world. Through our collective scientific endeavors, we aim to enhance our planet's

DOI: 10.1201/9781003423171-13

and its inhabitants' well-being. This chapter serves as a roadmap for the future of this field, offering guidance and insight into the imminent challenges. It calls the reader to action, encouraging them to consider biosecurity within synthetic biology.

13.1 ANTICIPATING FUTURE RESEARCH DIRECTIONS AND CHALLENGES

As synthetic biology continues to advance, new challenges and trends will undoubtedly emerge. Anticipating these can help researchers, policymakers, and the public prepare for the future and ensure that the benefits of synthetic biology are realized safely and ethically. Some challenges I anticipate in the near future include:

- **Increasing Complexity**: As synthetic biology techniques become more sophisticated, we are likely to see the creation of increasingly complex synthetic organisms. This raises new safety questions about how these organisms will behave and interact with natural ecosystems if they are accidentally or intentionally released.
- **Democratization of Biotechnology**: Advances in technology are making synthetic biology more accessible to a wider range of people. While this democratization of biotechnology has many benefits, it also poses challenges for oversight and control, as it may be harder to regulate and monitor synthetic biology techniques outside of traditional labs.
- **Use of Machine Learning and AI**: The use of machine learning and artificial intelligence in synthetic biology is expected to increase, which could accelerate the pace of discoveries and applications. However, integrating these technologies also presents new safety and ethical concerns, such as the potential for unintended consequences due to algorithmic bias or errors.
- **Climate Change and Biodiversity Loss**: Synthetic biology could play a role in addressing pressing global challenges such as climate change and biodiversity loss. For example, engineered organisms could capture carbon dioxide or protect endangered species. However, these applications also pose significant ecological risks that will need to be carefully managed.
- **Public Perception and Acceptance**: Public attitudes toward synthetic biology will continue to shape the field. Ongoing efforts will be needed to engage the public in dialogue about the benefits and risks of synthetic biology and to ensure that the development and application of this technology is guided by societal values and priorities.
- **International Collaboration and Regulation**: As synthetic biology becomes increasingly global, there will be a growing need for international collaboration and harmonization in certain crucial areas, such as safety standards, regulatory approaches, and response to potential biosecurity incidents.
- **Ethical and Social Implications**: As synthetic biology pushes the boundaries of what is possible, new ethical and social questions will arise. For example, the potential to create synthetic life or to genetically engineer humans poses profound questions about our understanding of life, identity, and what it means to be human.

13.2 LOOKING AHEAD: BALANCING INNOVATION AND SECURITY

Balancing innovation and security in synthetic biology requires thoughtful approaches to promoting scientific advancement while managing risks. Some practical ways this balance can be achieved include:

- **Regulation and Oversight**: Regulatory bodies at national and international levels need to be proactive in setting standards and guidelines for safe practices in synthetic biology. Regulations should be flexible enough to accommodate rapid scientific progress while maintaining rigorous safety and security measures. Clear pathways for regulatory approval can also help innovators to understand what is expected and avoid unnecessary obstacles.
- **Promote Safety by Design**: Incorporating safety and security considerations from the outset of designing a biological system or component can significantly mitigate risks. This approach integrates safety at every process stage – design, construction, testing, and deployment.
- **Dual-Use Research Policies**: Policies should be implemented for dual-use research, which has beneficial and potentially harmful applications. A system for reviewing and overseeing such research can ensure that potential risks are evaluated and mitigated while not stifling valuable scientific advancement.
- **Education and Training**: Providing scientists, bioengineers, and other practitioners with training in biosecurity and ethics can instill a culture of responsibility and awareness of potential risks. This includes undergraduate and graduate education, as well as professional development for those already in the field.
- **Transparency and Openness**: Open dialogue and transparent communication can help build public trust and enable the broader scientific community to assess and respond to potential risks. However, it is important to balance openness with the need to prevent misuse of information, especially dual-use research.
- **International Collaboration**: Biological risks are not confined to national borders. International cooperation is essential to share best practices, align safety standards, coordinate responses to incidents, and ensure wide access to the benefits of synthetic biology.
- **Public Engagement**: Public engagement plays a pivotal role in shaping both the perception and trajectory of this revolutionary technology. First, it educates the public about the nature of the technology, dispelling misconceptions and fostering a more nuanced understanding of its potential benefits and risks. This knowledge equips individuals to form informed opinions and participate in constructive dialogues about synthetic biology. Second, public engagement is key to maintaining transparency, essential for building trust between scientists, policymakers, and the public. This trust can be further enhanced by involving the public in decision-making processes related to synthetic biology, such as regulatory development, ethical considerations, and risk management strategies.

Moreover, public engagement can bring diverse perspectives to bear on the development and application of synthetic biology. These perspectives can shape research agendas, influence policy decisions, and highlight societal concerns that might be overlooked. Public involvement can also foster a sense of ownership and acceptance of synthetic biology and its applications. Various strategies can be employed to engage the public effectively. These might include town hall meetings, science festivals, educational programs in schools, citizen science projects, social media campaigns, and public consultations on policy decisions. Each approach has its unique strengths and can reach different segments of the public, promoting a wide-ranging and inclusive discussion about synthetic biology.

- **Ethical Guidelines and Codes of Conduct**: Professional organizations and institutions should establish ethical guidelines and codes of conduct that set clear expectations for responsible behavior in synthetic biology.

By integrating these practices, the synthetic biology community can ensure that innovation progresses. At the same time, potential risks are appropriately managed, contributing to the development of the field in a manner that is safe, ethical, and beneficial to society.

13.3 BIOSAFETY AND BIOSECURITY IN THE AGE OF SPACE EXPLORATION

As human civilization continues to push the boundaries of space exploration, the need for robust biorisk management is more pronounced than ever. As we aim to explore and potentially colonize other planets, several biosecurity and biosafety concerns emerge, which require careful contemplation and address:

- **Synthetic Biology in Space**: As the horizons of space exploration expand, the potential of leveraging synthetic biology in this context has become a topic of keen interest. Genetically engineered organisms could be used to produce essential materials, such as food, biofuels, medicines, or even terraforming extraterrestrial environments. However, this potential comes with considerable biorisk concerns. The behavior of engineered organisms in the space environment, vastly different from Earth, is unknown. There is a risk that these organisms could mutate, evolve, or proliferate in unpredictable ways under microgravity, radiation, or other space-specific factors. This could lead to unforeseen consequences, including potential threats to crew health or extraterrestrial ecosystems. Therefore, rigorous testing and containment strategies, as well as comprehensive risk assessments, are needed to harness synthetic biology in space safely.
- **Genetic Changes in Microorganisms**: Microorganisms have demonstrated an extraordinary ability to survive in the harsh conditions of space. Numerous studies have shown that bacteria and other microorganisms can undergo genetic changes in response to the unique stressors of space, such as high radiation levels and microgravity. These changes can potentially result in increased virulence (disease-causing ability) or enhanced antibiotic resistance

if these space-hardened microorganisms are returned to Earth; whether intentionally (as part of research) or unintentionally (on spacecraft), they could pose significant biosecurity and public health risks. Therefore, any mission involving the return of biological material from space must incorporate strict containment, sterilization, and quarantine protocols. Additionally, ongoing research is needed to understand the impact of space conditions on microbial behavior and genetics, enabling us to predict and mitigate these risks better.

- **Contamination of Other Planets (Forward Contamination)**: Before sending missions to other planets, it is crucial to consider the potential of contaminating these celestial bodies with Earth-based life forms. Despite stringent sterilization protocols, microorganisms have shown remarkable resilience in surviving extreme conditions. The inadvertent introduction of Earth's microorganisms could have unforeseen consequences on the alien ecosystem or confound the search for extraterrestrial life.
- **Back Contamination**: The prospect of bringing back extraterrestrial samples, whether geological or biological, poses the risk of "back contamination." If extraterrestrial life forms do exist, they may have adverse effects on Earth's ecosystem or pose direct risks to human health. It is crucial to have stringent quarantine and containment measures for handling and studying any returned samples.
- **Human Health in Space**: For crewed missions, maintaining human health and biosecurity is critical. Prolonged exposure to microgravity and radiation poses substantial health risks, including decreased immune response. Moreover, the closed environment of a spacecraft might allow microorganisms to proliferate and evolve in ways we do not yet fully understand, potentially leading to unexpected health risks.
- **Use of Synthetic Biology in Space**: Synthetic biology can be instrumental in space exploration, for instance, engineering organisms to produce food, fuel, or even terraforming a planet. However, these applications raise substantial biosecurity concerns. There is a risk of these genetically modified organisms escaping and potentially disrupting alien ecosystems.
- **Planetary Protection Policies**: To address these concerns, "Planetary Protection" policies have been created under the purview of international bodies like the Committee on Space Research (COSPAR).[177] These aim to prevent both forward and backward contamination during interplanetary travel. However, as our space exploration capabilities evolve, these policies would require continual updates.
- **International Collaboration**: Space exploration is a global endeavor and ensuring biosecurity and safety necessitates international collaboration. Shared protocols, transparency, and cooperation are essential for not only minimizing risks but also for effectively responding to any potential biosecurity events.

Addressing these concerns will require a multifaceted approach. Research is needed to understand how life (as we know it) behaves in space. Robust sterilization, containment, and quarantine procedures need to be developed and implemented. Most importantly, ethical guidelines and international regulations need to keep pace with the rapid advancements in our space exploration capabilities, ensuring that we explore responsibly and safely.

13.4 CYBERBIOSECURITY IN THE AGE OF SYNTHETIC BIOLOGY

Cyberbiosecurity broadly encompasses the understanding and mitigating of potential risks that emerge at the intersection of cybersecurity and biosecurity. In synthetic biology, this means ensuring the security, integrity, and confidentiality of biological data, systems, and technologies that are increasingly digitized and networked. Cyberbiosecurity is focused on protecting the life sciences landscape from threats such as unauthorized access, misuse of information, deliberate malicious acts, and accidental leaks that may compromise both cyber and biological systems.

- **Data Security in Synthetic Biology**: The advancements in synthetic biology have resulted in an explosion of data, ranging from genetic sequences to experimental procedures and proprietary intellectual property. As this data is increasingly digitized and shared across networks for research collaborations and open-source science, it becomes a potential target for cyber threats. Unauthorized access, data breaches, or manipulation could have serious consequences, including misinterpretation of data, incorrect experimental results, loss of proprietary information, and misuse of biological data for harmful purposes.
- **Security of Bioinformatics and Biotechnology Tools**: As synthetic biology relies heavily on bioinformatics tools, software, and automated laboratory equipment, ensuring their cybersecurity becomes essential. A breach in these systems could lead to sabotage of experiments, manipulation of results, or misuse of the equipment, potentially resulting in biohazardous situations. The Stuxnet attack, although not on a biological system, is a glaring example of how cyberattacks can manipulate machinery for destructive purposes.[178]
- **DNA Synthesis and Cyberbiosecurity**: DNA synthesis technologies, central to synthetic biology, pose a unique cyberbiosecurity concern. By providing the means to "write" any genetic sequence, they could potentially be exploited to synthesize harmful biological agents if the digital-to-biological conversion process is not appropriately safeguarded.
- **Regulatory Frameworks and Best Practices**: Addressing cyberbiosecurity requires comprehensive strategies that include strong regulatory frameworks, industry standards, and best practices for data security, encryption, authentication, access control, and system integrity checks. Policies should be dynamic and updated frequently to keep up with rapidly advancing technologies and emerging threats. Education and training programs for scientists, technicians, policymakers, and law enforcement must also integrate cyberbiosecurity aspects.
- **International Collaboration**: Cyberbiosecurity is a global concern, with threats capable of originating and impacting across borders. Therefore, international collaboration is paramount in sharing information about threats and best practices, as well as coordinating responses to global cyberbiosecurity incidents. This calls for greater dialogue and cooperation among nations, industries, and academia.

The interconnected world of synthetic biology opens up tremendous potential for advancements in medicine, agriculture, and environmental science, but it also introduces significant cyberbiosecurity challenges. Navigating this landscape requires robust, flexible, and globally coordinated efforts to ensure the security of our cyber and biological worlds as we harness the potential of synthetic biology.

13.5 AGRICULTURAL BIOSECURITY IN THE AGE OF SYNTHETIC BIOLOGY

Agricultural biosecurity is a set of measures designed to protect agriculture from biological threats such as pests, pathogens, invasive species, and genetically modified organisms. With the advent of synthetic biology, these measures are taking on new dimensions as the technology offers both opportunities and challenges for agricultural biosecurity:

- **Opportunities Offered by Synthetic Biology**: Synthetic biology can be leveraged to enhance agricultural biosecurity in numerous ways:
    - **Pathogen Detection and Monitoring**: Synthetic biology tools can develop novel diagnostics and biosensors for the rapid detection and monitoring of plant and animal diseases. For instance, gene-editing tools such as CRISPR have been used to develop biosensors that can detect specific viral pathogens in crops, aiding in early detection and swift containment of potential outbreaks.
    - **Development of Disease-Resistant Varieties**: Through genetic manipulation, synthetic biology can create crops and livestock resistant to diseases, pests, and environmental stresses, which can significantly reduce the dependence on pesticides and antibiotics and mitigate the spread of diseases.
    - **Biocontrol Agents**: Synthetic biology can engineer biocontrol agents that selectively target and eliminate specific pests or invasive species without harming other beneficial organisms or the environment.
- **Challenges Posed by Synthetic Biology**: While synthetic biology offers these opportunities, it also introduces new challenges to agricultural biosecurity:
    - **Dual-use Dilemma**: Synthetic biology techniques could potentially be misused to create harmful biological agents or to modify existing agricultural pathogens for malicious intent. This presents a dual-use dilemma and necessitates the development of stringent regulatory measures and ethical guidelines.
    - **Accidental Release**: There are concerns about the potential accidental release of genetically modified organisms into the environment, which could potentially disrupt ecosystems or crossbreed with wild relatives, leading to unpredictable consequences.
    - **Biosecurity Regulation**: Traditional agricultural biosecurity measures may be inadequate to deal with the risks associated with synthetic biology. This requires the development of updated biosecurity regulations that can effectively deal with the challenges posed by synthetic biology.

As we leverage the potential of synthetic biology for the advancement of agriculture, it is important that we navigate this new frontier with a vigilant eye on biosecurity. This calls for an anticipatory and integrated approach that considers both the exciting opportunities and the potential risks associated with synthetic biology and agricultural biosecurity.

13.6 GLOBAL CALL TO ACTION: ENSURING BIOSECURITY IN SYNTHETIC BIOLOGY

Ensuring biosecurity in the rapidly advancing field of synthetic biology is a global responsibility that requires commitment and collaborative efforts. Proposed "call to action" points that various stakeholders can consider include:

- **Invest in Safety and Security Research**: Governments and funding agencies should prioritize research that investigates the potential biosafety and biosecurity risks associated with synthetic biology, as well as the development of mitigation strategies.
- **Develop and Implement Policy Frameworks**: Policymakers at local, regional, national, and international levels should work together to develop and implement regulatory frameworks that ensure biosecurity, without hampering scientific progress.
- **Foster a Culture of Responsibility**: Institutions, organizations, and researchers involved in synthetic biology should foster a culture of responsibility, where safety, ethics, and security considerations are integrated into all aspects of the research process.
- **Promote Transparency and Open Dialogue**: It is crucial to encourage open dialogue about the potential risks and benefits of synthetic biology. This includes clear communication of research results and their implications, as well as an open discussion of potential misuse scenarios and mitigation strategies.
- **Strengthen Education and Training**: Countries should include biosecurity, biosafety, and bioethics in their established university curricula for life sciences and synthetic biology courses. This will equip the next generation of researchers with the knowledge and skills to navigate these issues.
- **Engage the Public**: Public engagement should be a priority to foster understanding and acceptance of synthetic biology and to ensure that societal values and concerns are considered in decision-making processes.
- **Establish and Enforce Codes of Conduct**: Professional organizations should establish codes of conduct for researchers working in synthetic biology, which explicitly state their responsibilities regarding safety and security.
- **Promote International Collaboration**: Ensuring biosecurity is a global issue that requires international cooperation. Countries should collaborate to share information, align regulatory approaches, and coordinate responses to biosecurity incidents.
- **Develop a Global Biosecurity Manual**: Developing a Global Biosecurity Manual will allow us to enhance and harmonize our ability to tackle and manage biosecurity effectively. This manual will provide standards and

guidelines based on biosecurity principles. By developing and adopting this manual, nations worldwide can collaboratively protect human and animal welfare, preserve biodiversity, and secure global well-being.

By taking these actions, we can ensure that synthetic biology advances in a way that is safe, responsible, and beneficial for all.

13.7 INDIVIDUAL CALL TO ACTION: ENSURING BIOSECURITY IN THE LAB

All researchers and their lab colleagues play a crucial role in maintaining biosafety and biosecurity in the era of synthetic biology. Some action items for the reader to undertake include:

13.7.1 Being Proactive in Biosecurity as a Synthetic Biologist

1. Understand and follow all relevant biorisk management regulations and guidelines in your country and institution.
2. Conduct thorough risk assessments before beginning any new projects and continue to assess risks as the project progresses.
3. Ensure that all safety equipment is properly maintained, and you and your colleagues are trained in its use.
4. Report any biosecurity incidents promptly and accurately; take steps to prevent similar incidents in the future.

13.7.2 Ensuring the Biorisk Management in Your Labs, Research, and Data

1. Use physical, biological, and procedural containment measures to prevent accidental release of synthetic organisms or other biohazards.
2. Implement security measures to protect sensitive data, such as encryption, access controls, and secure storage and transmission methods.
3. Maintain an inventory of your synthetic biology material and conduct regular inventory reconciliation to assess loss or theft.
4. Conduct regular audits of your biorisk management measures to identify and address any gaps or weaknesses.
5. Stay current with the latest advancements in biosafety and biosecurity and incorporate these into your practices.

13.7.3 Engaging in Continued Dialogue & Collaboration with Various Stakeholders

- **Colleagues**: Share best practices, discuss challenges, and collaborate on solutions. Encourage a culture of safety and security where everyone feels responsible for biosafety and biosecurity.

- **Scientific Community**: Participate in professional organizations, attend conferences, and contribute to scientific literature to learn from others and share your experiences.
- **Policymakers**: Inform policymakers about the importance of biorisk management in synthetic biology, and advocate for policies that support safe and responsible research.
- **Public**: Engage with the public to improve understanding of synthetic biology and its potential risks and benefits. Be transparent about your work and how you are managing risks.

By implementing these action items, researchers and their lab colleagues can contribute to a safer future in the age of synthetic biology.

13.8 CONCLUDING THOUGHTS

The emphasis on anticipating and preparing for future challenges in synthetic biology, a field marked by rapid advancements and complex ethical and biosecurity considerations, is significant. Adopting a forward-thinking mindset is necessary, and when combined with global collaboration and public engagement, it forms the cornerstone of responsible navigation through the complexities of this field. Such an approach ensures that synthetic biology's potential is harnessed beneficially for humanity while minimizing risks and ethical concerns. A profound and forward-looking perspective is necessary to ensure biosecurity in the rapidly evolving realm of synthetic biology. This perspective requires a comprehensive and collaborative approach involving many stakeholders, including scientists, policymakers, biorisk management experts, and the public. Pursuing scientific knowledge in this field must be coupled with responsible advancement, prioritizing biosecurity and ethical considerations. Anticipating future research directions and challenges is crucial to balancing innovation with biosecurity. Maintaining biosecurity in synthetic biology is a global responsibility that calls for concerted efforts, such as investing in safety and security research, developing and implementing policy frameworks, fostering a culture of accountability, and promoting international collaboration. Incorporating these strategies ensures that advancements in synthetic biology are innovative but also safe, ethical, and beneficial for society. The chapter's insights and recommendations provide a roadmap for the future, guiding the field toward responsible development and application. It also calls individual readers to take action on this vital topic. This roadmap is instrumental in shaping a future where synthetic biology contributes to societal progress with an unwavering commitment to safety, ethical integrity, and global cooperation.

13.9 KEY TAKEAWAYS

- **Comprehensive Approach**: Highlights the need for a comprehensive approach involving various stakeholders to ensure biosecurity.
- **Balancing Innovation and Security**: Focuses on the importance of balancing scientific advancement with security and ethical considerations.

- **Global Responsibility**: Stresses that maintaining biosecurity in synthetic biology is a global responsibility requiring collaborative efforts.
- **Anticipating Challenges**: Discusses the importance of anticipating future research directions and challenges in synthetic biology.
- **Promoting International Collaboration**: Emphasizes the need for international collaboration in addressing biosecurity concerns.
- **Fostering a Culture of Responsibility**: Advocates fostering a culture of responsibility and ethical consideration in synthetic biology research.
- **Public Engagement and Education**: Highlights the role of public engagement and education in shaping the future of synthetic biology.
- **A Call to Action**: The need for a global, comprehensive, and collaborative approach to ensure biosecurity in the field of synthetic biology is a current pressing issue that must be solved as soon as possible. A roadmap for ensuring that the advancements in synthetic biology are conducted in a manner that is safe, ethical, and beneficial for society at large should include the following elements:
    - **Involvement of Diverse Stakeholders**: Encouraging collaboration among scientists, policymakers, biorisk management experts, and the public to address the multifaceted challenges in synthetic biology.
    - **Balancing Innovation with Security**: Urging the synthetic biology community to balance the pursuit of scientific advancements with the need for biosecurity and ethical considerations.
    - **Global Responsibility**: Maintaining biosecurity in synthetic biology is a global responsibility, which requires concerted international efforts.
    - **Investing in Biosafety and Biosecurity Research**: Advocating for investment in research focused on biosafety and biosecurity to preemptively address potential risks associated with synthetic biology.
    - **Developing and Implementing Policy Frameworks**: Calling for the development and implementation of comprehensive policy frameworks to guide the ethical and secure advancement of synthetic biology.
    - **Fostering a Culture of Responsibility**: Promoting a culture of responsibility and ethical consideration within the synthetic biology research community.
    - **Promoting International Collaboration**: Emphasizing the importance of international collaboration in addressing biosecurity concerns and sharing best practices.
    - **Public Engagement and Education**: Stressing the role of public engagement and education in understanding and shaping the future of synthetic biology.

13.10 THOUGHT-PROVOKING QUESTIONS

13.1 What are the least understood factors for effectively managing potential risks associated with synthetic biology?

13.2 What should be updated or added to the current biorisk management regulatory framework so that the guidelines and policies can adequately handle the rapid advancements in synthetic biology?

13.3 What factors would you consider for developing a new model to predict the potential impacts of synthetic biology on nature's biodiversity?
13.4 What currently overlooked principles should be included in generating a new set of ethical guidelines for researchers in synthetic biology?
13.5 What factors should be included in designing a harmonized biosecurity framework/policy for space exploration?

Afterword Notes

As synthetic biology researchers push the frontier of knowledge and research, biosecurity will remain critical. This book provided specific guidelines for biosecurity risk assessment in various scenarios, from university labs to artificial intelligence. This book serves as a launch pad to consider the risks associated with the next generation of synthetic biology technologies that might be developed by the decade's end. Therefore, the biosafety and biosecurity community must be proactive, look to the future, find common ground, and pass international measures that address the biosafety and biosecurity of emerging subfields of synthetic biology. Some examples that were not covered in this book and should be covered in future publications are:

- **Astrobiology and the Next Phase of Development:** The creation of new or modified life forms designed to adapt and survive on other planets, such as Mars. Scientists have been running biological experiments at the international space station for many years, and the next natural step in the progression of that branch of science is to expand into creating *de novo* organisms that would be able to adapt and survive on other planets and the moon. We must consider the risks and consequences of such a line of experiments. What would happen if these new life forms were released on Earth? Also, what would be the risks and consequences of intentionally colonizing other planets or moons with newly created organisms? Should this be regulated? If so, by whom? Space exploration has recently become privatized. How should private spaceflight companies be regulated, and by whom? What about intentional or unintentional colonization of space or the moon by private citizens able to afford space travel for recreational purposes? Measures must be implemented to assess all the potential risks, reach consensus, and mitigate long before this capability becomes a reality.
- **Engineered Bacteria Designed to Purify Water and Food:** It is now possible to engineer and create bacteria capable of purifying food and water in the human digestive system. These bacteria would remove pollutants (heavy metals, chlorinated organics, "forever chemicals") from water and food before these pollutants are absorbed into the human body. These bacteria will "sequester" these pollutants and then be excreted. As these bacteria find their way into the ecosystem, what adverse effects will they have in the ecosystem? How would the local flora and fauna be affected? Humans travel all over the world. Therefore, a risk assessment would need to include the effects around every part of the world. It will become imperative to reach a consensus, assess all the risks, and implement mitigation measures using a "One Health" approach.
- **Engineered Bacteria Designed to Produce Painkillers or Give a High:** It is possible to engineer and create bacteria capable of living in our guts designed to produce painkillers or neurotransmitters that give humans a perennial state of happiness or ectasis. What would be the ethical

considerations for creating such organisms? What would be the potential environmental effects if the excreted bacteria colonized local animals, giving them the same effects? How could these risks be mitigated? A private biotech start-up can create such organisms or products completely outside any regulatory or legal framework, develop them, and market them simply as a natural probiotic supplement. How should this gap be mitigated? Would this product be considered a drug? The regulatory bodies need to anticipate this development and develop guidance and regulations to mitigate risks.

- **Longevity Treatments Potentially Being Abused:** New trends in the beauty and longevity spaces treat people with "stem cells," "exosomes," and a few other biologicals that fall outside the realm of FDA regulation. Undoubtedly, synthetic biology combined with the power of AI will bring about exponential development in this hotly popular topic. What measures must be in place to prevent the abuse of longevity "elixirs" that will be designed to counter arrest the cellular and muscular aging and decay of an 80-year-old person back 20 years being abused by teenagers wanting to stay forever young? What possible side effects do these treatments have on a developing brain and body? What irreversible damage could occur from being unnecessarily exposed to such treatments?
- **Computer Brain Interfaces (CBI):** In January 2024, the first brain chip developed by a private company was implanted into a person. The stated intention is to help people with paralysis or spinal cord injuries bypass the damaged neurons in their spine. However, this technology has the potential to be used for other reasons that may not be medically justified. There are many ethical concerns and questions regarding when and how these technologies interface the most inaccessible part of our body (the brain) with the internet. What sort of data could be commercialized and sold to others? Which privacy concerns should be addressed? What measures must be in place to prevent the hacking of someone's brain via a CBI?
- **Putting Synthetic Biology Elements on a Blockchain for Tracking:** As the international supply chain and payment systems move into these blockchain platforms, how will biological data be protected from threats of theft and use with nefarious intent? What elements of biosecurity should be included in a risk assessment? What mitigation measures would be needed to prevent tracking the orders a research lab has placed and gain knowledge of their research? How would this benefit and augment international IP theft? How far would the barrier to access be reduced? It is essential to continue to develop and bolster the new branch of biosecurity called cyberbiosecurity.

These issues are not brought up to alarm the readers, but rather to highlight that many areas of biotechnology research are rapidly evolving and changing dramatically and exponentially. We need to be proactive and not reactionary when these issues inevitably arrive.

– Anonymous Scientist

References

1. Chance, R. E.; Frank, B. H. Research, Development, Production, and Safety of Biosynthetic Human Insulin. *Diabetes Care* **1993**, *16*(Suppl 3), 133–142. https://doi.org/10.2337/diacare.16.3.133.
2. US EPA, EPA's Regulation of Bacillus thuringiensis(Bt) Crops | Pesticides | US EPA. https://www3.epa.gov/pesticides/chem_search/reg_actions/pip/regofbtcrops.htm (accessed 2023-07-19).
3. da Silva Fernandes, F.; de Souza, É. S.; Carneiro, L. M.; Alves Silva, J. P.; de Souza, J. V. B.; da Silva Batista, J. Current Ethanol Production Requirements for the Yeast Saccharomyces Cerevisiae. *Int. J. Microbiol.* **2022**, *2022*, 7878830. https://doi.org/10.1155/2022/7878830.
4. The Human Genome Project. Genome.gov. https://www.genome.gov/human-genome-project (accessed 2023-07-19).
5. Zhang, Y.; Cao, Y.; Luo, S.; Mukerabigwi, J. F.; Liu, M. Chapter 8 – Nanoparticles as Drug Delivery Systems of Combination Therapy for Cancer. In *Nanobiomaterials in Cancer Therapy*; Grumezescu, A. M., Ed. William Andrew Publishing: Norwich, NY, **2016**, pp 253–280. https://doi.org/10.1016/B978-0-323-42863-7.00008-6.
6. Mirtaleb, M. S.; Falak, R.; Heshmatnia, J.; Bakhshandeh, B.; Taheri, R. A.; Soleimanjahi, H.; Emameh, R. Z. An Insight Overview on COVID-19 mRNA Vaccines: Advantageous, Pharmacology, Mechanism of Action, and Prospective Considerations. *Int. Immunopharmacol.* **2023**, *117*, 109934. https://doi.org/10.1016/j.intimp.2023.109934.
7. Shinwari, Z. K.; Tanveer, F.; Khalil, A. T. Ethical Issues Regarding CRISPR Mediated Genome Editing. *Curr. Issues Mol. Biol.* **2018**, *26*, 103–110. https://doi.org/10.21775/cimb.026.103.
8. Mitochondrial Donation Treatment | HFEA. https://www.hfea.gov.uk/treatments/embryo-testing-and-treatments-for-disease/mitochondrial-donation-treatment/ (accessed 2023-07-19).
9. How Canadian Researchers Reconstituted an Extinct Poxvirus for $100,000 Using Mail-order DNA. https://www.science.org/content/article/how-canadian-researchers-reconstituted-extinct-poxvirus-100000-using-mail-order-dna (accessed 2023-07-19).
10. Dual Use Research of Concern and Bird Flu: Questions & Answers | Avian Influenza (Flu). https://www.cdc.gov/flu/avianflu/avian-durc-qa.htm (accessed 2023-07-19).
11. Research, C. E. 2001 Anthrax Attacks Fast Facts. *CNN.* https://www.cnn.com/2013/08/23/health/anthrax-fast-facts/index.html (accessed 2023-08-20).
12. Federal Select Agent Program. https://www.selectagents.gov/index.htm (accessed 2023-07-19).
13. Amerithrax or Anthrax Investigation. Federal Bureau of Investigation. https://www.fbi.gov/history/famous-cases/amerithrax-or-anthrax-investigation (accessed 2023-07-19).
14. Sohn, S.-I.; Pandian, S.; Oh, Y.-J.; Kang, H.-J.; Ryu, T.-H.; Cho, W.-S.; Shin, E.-K.; Shin, K.-S. A Review of the Unintentional Release of Feral Genetically Modified Rapeseed into the Environment. *Biology* **2021**, *10*(12), 1264. https://doi.org/10.3390/biology10121264.
15. GM Mosquito Progeny Not Dying in Brazil: Study. *The Scientist Magazine(r)*. https://www.the-scientist.com/news-opinion/gm-mosquito-progeny-not-dying-in-brazil--study-66434 (accessed 2023-07-19).
16. A Biotech CEO Explains Why He Injected Himself with a DIY Herpes Treatment on Facebook Live. *MIT Technology Review.* https://www.technologyreview.com/2018/02/05/145817/a-biotech-ceo-explains-why-he-injected-himself-with-a-diy-herpes-treatment-live-on-stage/ (accessed 2023-07-19).

17. Duroseau, B.; Kipshidze, N.; Limaye, R. J. The Impact of Delayed Access to COVID-19 Vaccines in Low- and Lower-Middle-Income Countries. *Front. Public Health* **2023**, *10*, 1087138. https://doi.org/10.3389/fpubh.2022.1087138.
18. Shahzad, M.; Faisal, M. S.; Shippey, E.; Divine, C.; Hoffman, M.; Mushtaq, M. U.; Shune, L.; McGuirk, J. P.; Ahmed, N. Socioeconomic and Racial Barriers to CD19 Chimeric Antigen Receptor T Cell Therapy (CART) Access. *Transplant. Cell. Ther.* **2022,** *28*(3), S208.
19. Badshah, S. L.; Ullah, A.; Ahmad, N.; Almarhoon, Z. M.; Mabkhot, Y. Increasing the Strength and Production of Artemisinin and Its Derivatives. *Mol. J. Synth. Chem. Nat. Prod. Chem.* **2018**, *23*(1), 100. https://doi.org/10.3390/molecules23010100.
20. Gibson, D. G.; Glass, J. I.; Lartigue, C.; Noskov, V. N.; Chuang, R.-Y.; Algire, M. A.; Benders, G. A.; Montague, M. G.; Ma, L.; Moodie, M. M.; Merryman, C.; Vashee, S.; Krishnakumar, R.; Assad-Garcia, N.; Andrews-Pfannkoch, C.; Denisova, E. A.; Young, L.; Qi, Z.-Q.; Segall-Shapiro, T. H.; Calvey, C. H.; Parmar, P. P.; Hutchison, C. A.; Smith, H. O.; Venter, J. C. Creation of a Bacterial Cell Controlled by a Chemically Synthesized Genome. *Science* **2010**, *329*(5987), 52–56. https://doi.org/10.1126/science.1190719.
21. CAR T Cells: Engineering Immune Cells to Treat Cancer – NCI. https://www.cancer.gov/about-cancer/treatment/research/car-t-cells (accessed 2023-07-19).
22. Hammond, A.; Galizi, R.; Kyrou, K.; Simoni, A.; Siniscalchi, C.; Katsanos, D.; Gribble, M.; Baker, D.; Marois, E.; Russell, S.; Burt, A.; Windbichler, N.; Crisanti, A.; Nolan, T. A CRISPR-Cas9 Gene Drive System Targeting Female Reproduction in the Malaria Mosquito Vector Anopheles Gambiae. *Nat. Biotechnol.* **2016**, *34* (1), 78–83. https://doi.org/10.1038/nbt.3439.
23. EXCLUSIVE: Chinese Scientists are Creating CRISPR Babies. *MIT Technology Review*. https://www.technologyreview.com/2018/11/25/138962/exclusive-chinese-scientists-are-creating-crispr-babies/ (accessed 2023-07-19).
24. OP#09: The 1971 Smallpox Epidemic in Aralsk, Kazakhstan, and the Soviet Biological Warfare Program. *James Martin Center for Nonproliferation Studies*. https://nonproliferation.org/the-1971-smallpox-epidemic-in-aralsk-kazakhstan-and-the-soviet-biological-warfare-program-commentaries/ (accessed 2023-08-20).
25. Aralsk: A Kazakh Town That Lived through a Smallpox Epidemic. Davis Center. https://daviscenter.fas.harvard.edu/insights/aralsk-kazakh-town-lived-through-smallpox-epidemic (accessed 2023-08-01).
26. Anthrax at Sverdlovsk. **1979**. https://nsarchive2.gwu.edu/NSAEBB/NSAEBB61/ (accessed 2023-08-20).
27. Meselson, M.; Guillemin, J.; Hugh-Jones, M.; Langmuir, A.; Popova, I.; Shelokov, A.; Yampolskaya, O. The Sverdlovsk Anthrax Outbreak of 1979. *Science* **1994**, *266*(5188), 1202–1208. https://doi.org/10.1126/science.7973702.
28. Reports Blame Lab for Foot-and-Mouth Fiasco. https://www.science.org/content/article/reports-blame-lab-foot-and-mouth-fiasco (accessed 2023-08-22).
29. Foot and Mouth Disease 2007: A Review and Lessons Learned. GOV.UK. https://www.gov.uk/government/publications/foot-and-mouth-disease-2007-a-review-and-lessons-learned (accessed 2023-08-01).
30. Gronvall, G. K. H5N1: A Case Study for Dual-Use Research. **2013**. https://cdn.cfr.org/sites/default/files/pdf/2013/05/WP_Dual_Use_Research.pdf
31. EXCLUSIVE: Controversial Experiments That Could Make Bird Flu More Risky Poised to Resume. https://www.science.org/content/article/exclusive-controversial-experiments-make-bird-flu-more-risky-poised-resume (accessed 2023-08-01).
32. Lancet, T. The COVID-19 Pandemic in 2023: Far from Over. *The Lancet* **2023**, *401*(10371), 79. https://doi.org/10.1016/S0140-6736(23)00050-8.
33. Josefson, D. Scientists Manage to Manufacture Polio Virus. *BMJ* **2002**, *325*(7356), 122.

34. Knight, T. Idempotent Vector Design for Standard Assembly of Biobricks. MIT Artificial Intelligence Laboratory; MIT Synthetic Biology Working Group. **2003**. http://hdl.handle.net/1721.1/21168
35. Ro, D.-K.; Paradise, E. M.; Ouellet, M.; Fisher, K. J.; Newman, K. L.; Ndungu, J. M.; Ho, K. A.; Eachus, R. A.; Ham, T. S.; Kirby, J.; Chang, M. C. Y.; Withers, S. T.; Shiba, Y.; Sarpong, R.; Keasling, J. D. Production of the Antimalarial Drug Precursor Artemisinic Acid in Engineered Yeast. *Nature* **2006**, *440*(7086), 940–943. https://doi.org/10.1038/nature04640.
36. BioBricks Foundation | Biotechnology in the Public Interest. https://biobricks.org/ (accessed 2023-11-06).
37. Gardner, T. S.; Cantor, C. R.; Collins, J. J. Construction of a Genetic Toggle Switch in Escherichia Coli. *Nature* **2000**, *403*(6767), 339–342. https://doi.org/10.1038/35002131.
38. Jinek, M.; Chylinski, K.; Fonfara, I.; Hauer, M.; Doudna, J. A.; Charpentier, E. A Programmable Dual-RNA-Guided DNA Endonuclease in Adaptive Bacterial Immunity. *Science* **2012**, *337*(6096), 816–821. https://doi.org/10.1126/science.1225829.
39. Richards, D. J.; Tan, Y.; Jia, J.; Yao, H.; Mei, Y. 3D Printing for Tissue Engineering. *Isr. J. Chem.* **2013**, *53*(9–10), 805–814.
40. Malonis, R. J.; Lai, J. R.; Vergnolle, O. Peptide-Based Vaccines: Current Progress and Future Challenges. *Chem. Rev.* **2020**, *120*(6), 3210–3229. https://doi.org/10.1021/acs.chemrev.9b00472.
41. The Nobel Prize in Physiology or Medicine 2015. NobelPrize.org. https://www.nobelprize.org/prizes/medicine/2015/tu/facts/ (accessed 2023-11-06).
42. Salerno, R. M.; Gaudioso, J., Eds. *Laboratory Biorisk Management: Biosafety and Biosecurity*, 1st edition. CRC Press: Boca Raton, FL, **2015**.
43. Hutchison, C. A.; Chuang, R.-Y.; Noskov, V. N.; Assad-Garcia, N.; Deerinck, T. J.; Ellisman, M. H.; Gill, J.; Kannan, K.; Karas, B. J.; Ma, L.; Pelletier, J. F.; Qi, Z.-Q.; Richter, R. A.; Strychalski, E. A.; Sun, L.; Suzuki, Y.; Tsvetanova, B.; Wise, K. S.; Smith, H. O.; Glass, J. I.; Merryman, C.; Gibson, D. G.; Venter, J. C. Design and Synthesis of a Minimal Bacterial Genome. *Science* **2016**, *351*(6280), aad6253. https://doi.org/10.1126/science.aad6253.
44. Wu, Y.; Liu, Y.; Wang, T.; Jiang, Q.; Xu, F.; Liu, Z. Living Cell for Drug Delivery. *Eng. Regen.* **2022**, *3*(2), 131–148. https://doi.org/10.1016/j.engreg.2022.03.001.
45. Bier, E. Gene Drives Gaining Speed. *Nat. Rev. Genet.* **2022**, *23*(1), 5–22. https://doi.org/10.1038/s41576-021-00386-0.
46. The Nobel Prize in Chemistry 2016. NobelPrize.org. https://www.nobelprize.org/prizes/chemistry/2016/summary/ (accessed 2023-11-06).
47. Zhang, Y.; Ptacin, J. L.; Fischer, E. C.; Aerni, H. R.; Caffaro, C. E.; San Jose, K.; Feldman, A. W.; Turner, C. R.; Romesberg, F. E. A Semi-Synthetic Organism That Stores and Retrieves Increased Genetic Information. *Nature* **2017**, *551*(7682), 644–647. https://doi.org/10.1038/nature24659.
48. Richardson, S. M.; Mitchell, L. A.; Stracquadanio, G.; Yang, K.; Dymond, J. S.; DiCarlo, J. E.; Lee, D.; Huang, C. L. V.; Chandrasegaran, S.; Cai, Y.; Boeke, J. D.; Bader, J. S. Design of a Synthetic Yeast Genome. Science **2017**, 355(6329), 1040–1044. https://doi.org/10.1126/science.aaf4557.
49. The Nobel Prize in Chemistry 2018. NobelPrize.org. https://www.nobelprize.org/prizes/chemistry/2018/arnold/facts/ (accessed 2023-11-06).
50. Living Machines Are Created in the Lab | Tufts Now. https://now.tufts.edu/2020/01/16/living-machines-are-created-lab (accessed 2023-11-06).
51. Kriegman, S.; Blackiston, D.; Levin, M.; Bongard, J. A Scalable Pipeline for Designing Reconfigurable Organisms. *Proc. Natl. Acad. Sci.* **2020**, *117*(4), 1853–1859. https://doi.org/10.1073/pnas.1910837117.
52. The Nobel Prize in Chemistry 2020. NobelPrize.org. https://www.nobelprize.org/prizes/chemistry/2020/summary/ (accessed 2023-11-05).

53. In a First, Scientists Create Tiny Multicellular Organisms That Can Replicate | Tufts Now. https://now.tufts.edu/2021/11/29/first-scientists-create-tiny-multicellular-organisms-can-replicate (accessed 2023-11-06).
54. Bacterial Biosensors: The Future of Analyte Detection. ASM.org. https://asm.org:443/Articles/2023/September/Bacterial-Biosensors-The-Future-of-Analyte-Detecti (accessed 2023-11-06).
55. AI Technology Generates Original Proteins from Scratch | UC San Francisco. https://www.ucsf.edu/news/2023/01/424641/ai-technology-generates-original-proteins-scratch (accessed 2023-11-06).
56. In a World First, Vertex, CRISPR Win UK Approval for CRISPR-edited Therapy to Treat Sickle Cell Disease, Beta-thalassemia. *Endpoints News*. https://endpts.com/uk-approves-vertex-crispr-therapy-for-sickle-cell-disease-beta-thalassemia-in-world-first/ (accessed 2023-11-17).
57. The Nobel Prize in Physiology or Medicine 2023. NobelPrize.org. https://www.nobelprize.org/prizes/medicine/2023/summary/ (accessed 2023-10-25).
58. iGEM Video Universe. https://video.igem.org/videos/embed/1cfcd8d9-9a4e-42ae-9451-21267b4860b4?title=0&warningTitle=0?warningTitle=0&peertubeLink=0 (accessed 2023-07-19).
59. Ginkgo Bioworks. *Wikipedia*, **2023**.
60. Wu, F.; Wesseler, J.; Zilberman, D.; Russell, R. M.; Chen, C.; Dubock, A. C. Allow Golden Rice to Save Lives. *Proc. Natl. Acad. Sci.* **2021**, *118*(51), e2120901118. https://doi.org/10.1073/pnas.2120901118.
61. Adamala, K.; Szostak, J. W. Nonenzymatic Template-Directed RNA Synthesis Inside Model Protocells. *Science* **2013**, *342*(6162), 1098–1100. https://doi.org/10.1126/science.1241888.
62. Zamani, M.; Berenjian, A.; Hemmati, S.; Nezafat, N.; Ghoshoon, M. B.; Dabbagh, F.; Mohkam, M.; Ghasemi, Y. Cloning, Expression, and Purification of a Synthetic Human Growth Hormone in Escherichia Coli Using Response Surface Methodology. *Mol. Biotechnol.* **2015**, *57*(3), 241–250. https://doi.org/10.1007/s12033-014-9818-1.
63. Kan, S. B. J.; Lewis, R. D.; Chen, K.; Arnold, F. H. Directed Evolution of Cytochrome c for Carbon-Silicon Bond Formation: Bringing Silicon to Life. *Science* **2016**, *354*(6315), 1048–1051. https://doi.org/10.1126/science.aah6219.
64. Gote, V.; Bolla, P. K.; Kommineni, N.; Butreddy, A.; Nukala, P. K.; Palakurthi, S. S.; Khan, W. A Comprehensive Review of mRNA Vaccines. *Int. J. Mol. Sci.* **2023**, *24*(3), 2700. https://doi.org/10.3390/ijms24032700.
65. Frangoul, H.; Altshuler, D.; Cappellini, M. D.; Chen, Y.-S.; Domm, J.; Eustace, B. K.; Foell, J.; de la Fuente, J.; Grupp, S.; Handgretinger, R.; Ho, T. W.; Kattamis, A.; Kernytsky, A.; Lekstrom-Himes, J.; Li, A. M.; Locatelli, F.; Mapara, M. Y.; de Montalembert, M.; Rondelli, D.; Sharma, A.; Sheth, S.; Soni, S.; Steinberg, M. H.; Wall, D.; Yen, A.; Corbacioglu, S. CRISPR-Cas9 Gene Editing for Sickle Cell Disease and β-Thalassemia. *N. Engl. J. Med.* **2021**, *384*(3), 252–260. https://doi.org/10.1056/NEJMoa2031054.
66. Tissue Printer Creates Lifelike Human Ear. https://www.science.org/content/article/tissue-printer-creates-lifelike-human-ear (accessed 2023-07-20).
67. "Nanoemulsion" Gels Offer New Way to Deliver Drugs through the Skin. *MIT News | Massachusetts Institute of Technology*. https://news.mit.edu/2019/nanoemulsion-gels-skin-drugs-0621 (accessed 2023-07-20).
68. Xu, J.; Wen, Z. Brain Organoids: Studying Human Brain Development and Diseases in a Dish. *Stem Cells Int.* **2021**, *2021*, 5902824. https://doi.org/10.1155/2021/5902824.
69. Lung-on-a-Chip. Wyss Institute. https://wyss.harvard.edu/media-post/lung-on-a-chip/ (accessed 2023-07-20).
70. Abulaiti, M.; Yalikun, Y.; Murata, K.; Sato, A.; Sami, M. M.; Sasaki, Y.; Fujiwara, Y.; Minatoya, K.; Shiba, Y.; Tanaka, Y.; Masumoto, H. Establishment of a Heart-on-a-Chip Microdevice Based on Human iPS Cells for the Evaluation of Human Heart Tissue Function. *Sci. Rep.* **2020**, *10*(1), 19201. https://doi.org/10.1038/s41598-020-76062-w.

71. Chani, B.; Puri, V.; Sobti, R. C.; Jha, V.; Puri, S. Decellularized Scaffold of Cryopreserved Rat Kidney Retains Its Recellularization Potential. *PLoS One* **2017**, *12*(3), e0173040. https://doi.org/10.1371/journal.pone.0173040.
72. Noyce, R. S.; Lederman, S.; Evans, D. H. Construction of an Infectious Horsepox Virus Vaccine from Chemically Synthesized DNA Fragments. *PLoS One* **2018**, *13*(1), e0188453. https://doi.org/10.1371/journal.pone.0188453.
73. Mustafa, M. I.; Makhawi, A. M. SHERLOCK and DETECTR: CRISPR-Cas Systems as Potential Rapid Diagnostic Tools for Emerging Infectious Diseases. *J. Clin.* Microbiol. **2021**, *59*(3). https://doi.org/10.1128/jcm.00745-20.
74. A New Company with a Wild Mission: Bring Back the Woolly Mammoth. *The New York Times*. https://www.nytimes.com/2021/09/13/science/colossal-woolly-mammoth-DNA.html (accessed 2023-07-22).
75. USDA APHIS | Biosecurity and Other Programs. https://www.aphis.usda.gov/ (accessed 2023-08-22).
76. Biosecurity | Biosecurity. https://www.biosecurity.gov.au/ (accessed 2023-08-22).
77. Ministry for Primary Industries. Biosecurity | MPI – Ministry for Primary Industries. A New Zealand Government Department. https://www.mpi.govt.nz/biosecurity/ (accessed 2023-08-22).
78. Kaiser, J. Resurrected Influenza Virus Yields Secrets of Deadly 1918 Pandemic. *Science* **2005**, *310*(5745), 28–29. https://doi.org/10.1126/science.310.5745.28.
79. Bayer, M. **2023**. *Big Pharma-Partnered Evotec on High Alert after Cyberattack Takes Systems Offline*. Fierce Biotech. https://www.fiercebiotech.com/biotech/big-pharma-partnered-evotec-closes-down-hatchets-after-cyber-attack (accessed 2023-07-23).
80. Schumacher, G. J.; Sawaya, S.; Nelson, D.; Hansen, A. J. Genetic Information Insecurity as State of the Art. *Front. Bioeng. Biotechnol.* **2020**, *8*, 591980. https://doi.org/10.3389/fbioe.2020.591980.
81. Li, J.; Xiao, W.; Zhang, C. Data Security Crisis in Universities: Identification of Key Factors Affecting Data Breach Incidents. *Humanit. Soc. Sci. Commun.* **2023**, *10*(1), 1–18. https://doi.org/10.1057/s41599-023-01757-0.
82. Characterization of the Reconstructed 1918 Spanish Influenza Pandemic Virus | Science. https://www.science.org/doi/10.1126/science.1119392 (accessed 2023-07-19).
83. Flachsbart, F.; Caliebe, A.; Kleindorp, R.; Blanché, H.; von Eller-Eberstein, H.; Nikolaus, S.; Schreiber, S.; Nebel, A. Association of FOXO3A Variation with Human Longevity Confirmed in German Centenarians. *Proc. Natl. Acad. Sci. U. S. A.* **2009**, *106*(8), 2700–2705. https://doi.org/10.1073/pnas.0809594106.
84. Jumper, J.; Evans, R.; Pritzel, A.; Green, T.; Figurnov, M.; Ronneberger, O.; Tunyasuvunakool, K.; Bates, R.; Žídek, A.; Potapenko, A.; Bridgland, A.; Meyer, C.; Kohl, S. A. A.; Ballard, A. J.; Cowie, A.; Romera-Paredes, B.; Nikolov, S.; Jain, R.; Adler, J.; Back, T.; Petersen, S.; Reiman, D.; Clancy, E.; Zielinski, M.; Steinegger, M.; Pacholska, M.; Berghammer, T.; Bodenstein, S.; Silver, D.; Vinyals, O.; Senior, A. W.; Kavukcuoglu, K.; Kohli, P.; Hassabis, D. Highly Accurate Protein Structure Prediction with AlphaFold. *Nature* **2021**, *596*(7873), 583–589. https://doi.org/10.1038/s41586-021-03819-2.
85. Savransky, V.; Shearer, J. D.; Gainey, M. R.; Sanford, D. C.; Sivko, G. S.; Stark, G. V.; Li, N.; Ionin, B.; Lacy, M. J.; Skiadopoulos, M. H. Correlation between Anthrax Lethal Toxin Neutralizing Antibody Levels and Survival in Guinea Pigs and Nonhuman Primates Vaccinated with the AV7909 Anthrax Vaccine Candidate. *Vaccine* **2017**, *35*(37), 4952–4959. https://doi.org/10.1016/j.vaccine.2017.07.076.
86. House, T. W. Executive Order on the Safe, Secure, and Trustworthy Development and Use of Artificial Intelligence. *The White House*. https://www.whitehouse.gov/briefing-room/presidential-actions/2023/10/30/executive-order-on-the-safe-secure-and-trustworthy-development-and-use-of-artificial-intelligence/ (accessed 2023-11-26).

87. Pandemic Preparedness in a Changing World: Fostering Global Collaboration to Strengthen Public Health and Response to Viral Threats. https://www.science.org/content/resource/pandemic-preparedness-changing-world-fostering-global-collaboration (accessed 2023-08-20).
88. Maxmen, A. Opinion | Why Isn't the U.S. Embracing This Pandemic Prevention Strategy? *The New York Times*. December 6, **2022**. https://www.nytimes.com/2022/12/06/opinion/us-covid-vaccines-pandemic-prevention.html (accessed 2023-08-20).
89. Unseating Big Pharma: The Radical Plan for Vaccine Equity. https://www.nature.com/immersive/d41586-022-01898-3/index.html (accessed 2023-08-20).
90. De Haro, L. P. Biosecurity Risk Assessment for the Use of Artificial Intelligence in Synthetic Biology. *Applied Biosafety* **2024** *29*(2). DOI: 10.1089/apb.2023.00311; https://www.liebertpub.com/doi/full/10.1089/apb.2023.0031
91. Maserat, E. Integration of Artificial Intelligence and CRISPR/Cas9 System for Vaccine Design. *Cancer Inform.* **2022**, *21*, 11769351221140102. https://doi.org/10.1177/11769351221140102.
92. Méndez-Lucio, O.; Baillif, B.; Clevert, D.-A.; Rouquié, D.; Wichard, J. De Novo Generation of Hit-like Molecules from Gene Expression Signatures Using Artificial Intelligence. *Nat. Commun.* **2020**, *11*(1), 10. https://doi.org/10.1038/s41467-019-13807-w.
93. Ichikawa, D. M.; Abdin, O.; Alerasool, N.; Kogenaru, M.; Mueller, A. L.; Wen, H.; Giganti, D. O.; Goldberg, G. W.; Adams, S.; Spencer, J. M.; Razavi, R.; Nim, S.; Zheng, H.; Gionco, C.; Clark, F. T.; Strokach, A.; Hughes, T. R.; Lionnet, T.; Taipale, M.; Kim, P. M.; Noyes, M. B. A Universal Deep-Learning Model for Zinc Finger Design Enables Transcription Factor Reprogramming. *Nat. Biotechnol.* **2023**, *41*(8), 1117–1129. https://doi.org/10.1038/s41587-022-01624-4
94. Eisenstein, M. AI-Enhanced Protein Design Makes Proteins That Have Never Existed. *Nat. Biotechnol.* **2023**, *41*(3), 303–305. https://doi.org/10.1038/s41587-023-01705-y.
95. API. Wikipedia; **2024**.
96. Sharma, A.; Virmani, T.; Pathak, V.; Sharma, A.; Pathak, K.; Kumar, G.; Pathak, D. Artificial Intelligence-Based Data-Driven Strategy to Accelerate Research, Development, and Clinical Trials of COVID Vaccine. *BioMed Res. Int.* **2022**, *2022*, 7205241. https://doi.org/10.1155/2022/7205241.
97. Prasad, K.; Cross, R. S.; Jenkins, M. R. Synthetic Biology, Genetic Circuits and Machine Learning: A New Age of Cancer Therapy. *Mol. Oncol.* **2023**, *17*(6), 946–949. https://doi.org/10.1002/1878-0261.13420.
98. Radivojević, T.; Costello, Z.; Workman, K.; Garcia Martin, H. A Machine Learning Automated Recommendation Tool for Synthetic Biology. *Nat. Commun.* **2020**, *11*(1), 4879. https://doi.org/10.1038/s41467-020-18008-4.
99. Callaway, E. How Generative AI Is Building Better Antibodies. *Nature* **2023**, *617*(7960), 235–235. https://doi.org/10.1038/d41586-023-01516-w.
100. Gentile, F.; Yaacoub, J. C.; Gleave, J.; Fernandez, M.; Ton, A.-T.; Ban, F.; Stern, A.; Cherkasov, A. Artificial Intelligence-Enabled Virtual Screening of Ultra-Large Chemical Libraries with Deep Docking. *Nat. Protoc.* **2022**, *17*(3), 672–697. https://doi.org/10.1038/s41596-021-00659-2.
101. Autonomous Discovery | Argonne National Laboratory. https://www.anl.gov/autonomous-discovery (accessed 2023-11-25).
102. Ziller, A.; Usynin, D.; Braren, R.; Makowski, M.; Rueckert, D.; Kaissis, G. Medical Imaging Deep Learning with Differential Privacy. *Sci. Rep.* **2021**, *11*(1), 13524. https://doi.org/10.1038/s41598-021-93030-0.
103. Cheng, Y.; Liu, Y.; Chen, T.; Yang, Q. Federated Learning for Privacy-Preserving AI. *Commun. ACM* **2020**, *63*(12), 33–36. https://doi.org/10.1145/3387107.
104. Alabdulatif, A.; Khalil, I.; Saidur Rahman, M. Security of Blockchain and AI-Empowered Smart Healthcare: Application-Based Analysis. *Appl. Sci.* **2022**, *12*(21), 11039. https://doi.org/10.3390/app122111039.

105. Rudin, C.; Radin, J. Why Are We Using Black Box Models in AI When We Don't Need To? A Lesson from an Explainable AI Competition. *Harv. Data Sci. Rev.* **2019**, 1(2). https://doi.org/10.1162/99608f92.5a8a3a3d.
106. Rudin, C. Stop Explaining Black Box Machine Learning Models for High Stakes Decisions and Use Interpretable Models Instead. *Nat. Mach. Intell.* **2019**, *1*(5), 206–215. https://doi.org/10.1038/s42256-019-0048-x.
107. Myllyaho, L.; Raatikainen, M.; Männistö, T.; Mikkonen, T.; Nurminen, J. K. Systematic Literature Review of Validation Methods for AI Systems. *J. Syst. Softw.* **2021**, *181*, 111050. https://doi.org/10.1016/j.jss.2021.111050.
108. Foss-Solbrekk, K. Three Routes to Protecting AI Systems and Their Algorithms under IP Law: The Good, the Bad and the Ugly. *J. Intellect. Prop. Law Pract.* **2021**, *16*(3), 247–258. https://doi.org/10.1093/jiplp/jpab033.
109. Hartmann, K.; Steup, C. Hacking the AI: The Next Generation of Hijacked Systems. In *2020 12th International Conference on Cyber Conflict (CyCon);* **2020**; Vol. 1300, pp 327–349. https://doi.org/10.23919/CyCon49761.2020.9131724.
110. Steimers, A.; Schneider, M. Sources of Risk of AI Systems. *Int. J. Environ. Res. Public. Health* **2022**, *19*(6), 3641. https://doi.org/10.3390/ijerph19063641.
111. O'Brien, J. T.; Nelson, C. Assessing the Risks Posed by the Convergence of Artificial Intelligence and Biotechnology. *Health Secur.* **2020**, *18*(3), 219–227. https://doi.org/10.1089/hs.2019.0122.
112. 14:00-17:00. ISO/IEC 22989:2022. ISO. https://www.iso.org/standard/74296.html (accessed 2023-11-23).
113. Sandbrink, J. B. Artificial Intelligence and Biological Misuse: Differentiating Risks of Language Models and Biological Design Tools. https://arxiv.org/abs/2306.13952
114. Biological Weapons Convention. **1975**. https://disarmament.unoda.org/biological-weapons/ (accessed 2023-07-16).
115. The Biological Weapons Convention (BWC) At A Glance | Arms Control Association. https://www.armscontrol.org/factsheets/bwc (accessed 2023-08-23).
116. Cartagena Protocol on Biosafety to the Convention on Biological Diversity. **2000**. https://treaties.un.org/Pages/ViewDetails.aspx?src=IND&mtdsg_no=XXVII-8-a&chapter=27&clang=_en.
117. Biosecurity & Health Security Protection (BSP), Viet Nam. Nagoya Protocol on Access to Genetic Resources and the Fair and Equitable Sharing of Benefits Arising from Their Utilization to the Convention on Biological Diversity. World Health Organization (editors) **2010**. https://www.who.int/publications/i/item/9789240011311.
118. *Laboratory Biosafety Manual*, 4th edition. WHO, **2020**.
119. 18 USC Ch. 10: Biological Weapons. https://uscode.house.gov/view.xhtml?path=/prelim@title18/part1/chapter10&edition=prelim (accessed 2023-11-21).
120. DOD Unveils Collaborative Biodefense Reforms in Posture Review. U.S. Department of Defense. https://www.defense.gov/News/Releases/Release/Article/3495836/dod-unveils-collaborative-biodefense-reforms-in-posture-review/#:~:text=%22The%20Biodefense%20Posture%20Review%20fully,global%20risk%20of%20laboratory%20accidents.%22 (accessed 2023-08-22).
121. "Fairly Shocking": Secret Medical Lab in California Stored Bioengineered Mice Laden with COVID. *USA Today*. https://www.usatoday.com/story/news/nation/2023/07/31/illegal-lab-california-infectious-mice/70502532007/ (accessed 2023-08-27).
122. Meechan, P. J.; Potts, J. *Biosafety in Microbiological and Biomedical Laboratories*, 6th edition. CDC & NIH, **2020**. https://bioethicsarchive.georgetown.edu/pcsbi/synthetic-biology-report.html
123. New Directions: The Ethics of Synthetic Biology and Emerging Technologies, **2012**.
124. NIH Guidelines for Research Involving Recombinant or Synthetic Nucleic Acid Molecules (NIH Guidelines), April 2019. **2019**. https://osp.od.nih.gov/wp-content/uploads/NIH_Guidelines.pdf

125. Read "Biodefense in the Age of Synthetic Biology" at NAP.Edu. https://doi.org/10.17226/24890.
126. Field, M. Biotech Promises Miracles. But the Risks Call for more Oversight. *Bulletin of the Atomic Scientists*. https://thebulletin.org/2023/08/biotech-promises-miracles-but-the-risks-call-for-more-oversight/ (accessed 2023-09-24).
127. Government of Canada. *Canadian Biosafety Standard, Third Edition*. https://www.canada.ca/en/public-health/services/canadian-biosafety-standards-guidelines/third-edition.html (accessed 2023-07-16).
128. Government of Canada. *Canadian Biosafety Handbook, Second Edition*. https://www.canada.ca/en/public-health/services/canadian-biosafety-standards-guidelines/handbook-second-edition.html (accessed 2024-02-04).
129. PHAC Training Portal. https://training-formation.phac-aspc.gc.ca/?lang=en (accessed 2024-02-04).
130. Government of Canada. *Pathogen Safety Data Sheets*. https://www.canada.ca/en/public-health/services/laboratory-biosafety-biosecurity/pathogen-safety-data-sheets-risk-assessment.html (accessed 2024-02-04).
131. European Biosafety Network (EBN). European Biosafety Network. https://www.europeanbiosafetynetwork.eu/ (accessed 2023-07-16).
132. Plant Health and Biosecurity. https://food.ec.europa.eu/plants/plant-health-and-biosecurity_en (accessed 2023-07-16).
133. Africa CDC Biosafety and Biosecurity Initiative Report on the Consultative Process to Identify Priorities for Strengthening Biosafety and Biosecurity. https://biosecuritycentral.org/resource/training-materials/africa-cdc-bsbs-report/ (accessed 2023-07-16).
134. Biosafety and Biosecurity Initiative 2021–2025 Strategic Plan. *Africa CDC*. https://africacdc.org/download/biosafety-and-biosecurity-initiative-2021-2025-strategic-plan/ (accessed 2023-07-16).
135. Biosecurity Act 2015. Federal Register of Legislation of the Australian Government; **2023**. https://www.legislation.gov.au/C2015A00061/latest/text
136. Agriculture. Biosecurity Bill 2014. https://www.legislation.gov.au/Details/C2014B00245/Html/Text, https://www.legislation.gov.au/Details/C2014B00245 (accessed 2023-07-16).
137. Southeast Asia Strategic Biosecurity Dialogue | Johns Hopkins Center for Health Security. https://centerforhealthsecurity.org/our-work/research-projects/southeast-asia-strategic-biosecurity-dialogue (accessed 2023-07-16).
138. Wang, L.; Song, J.; Zhang W. Tianjin Biosecurity Guidelines for codes of conduct for scientists: Promoting responsible sciences and strengthening biosecurity governance. *J Biosaf. Biosecurity* **2021**, *3*, 82–83. https://doi.org/10.1016/j.jobb.2021.08.001
139. Hamilton, R. A. Biological Security Priorities in the Middle East_2020. https://unicri.it/sites/default/files/2020-11/Middle%20East_0.pdf
140. Khan, E.; Ahmed, N.; Temsamani, K. R.; El-Gendy, A.; Cohen, M.; Hasan, A.; Gastfriend, H.; Cole, J. Biosafety Initiatives in BMENA Region: Identification of Gaps and Advances. *Front. Public Health* **2016**, *4*. https://doi.org/10.3389/fpubh.2016.00044.
141. Elizeus, R.; Fortress, Y., Aku; Eva, I., Kayed, Al, Zein; Hedia, B. Reasons for and Barriers to Biosafety and Biosecurity Training in Health-Related Organizations in Africa, Middle East and Central Asia: Findings from GIBACHT Training Needs Assessments 2018-2019. *PAMJ* **2020**, *37*(64). https://doi.org/10.11604/pamj.2020.37.64.23390.
142. Sociedade Brasileira de Biossegurança e Bioproteção. https://sb3.org.br/ (accessed 2023-07-16).
143. Indian Laws & Regulations – ICAR Biosafety Portal. https://biosafety.icar.gov.in/category/indianlawsandregulations/ (accessed 2023-07-16).
144. The Agricultural Biosecurity Bill, **2013**. PRS Legislative Research. https://prsindia.org/billtrack/the-agricultural-biosecurity-bill-2013 (accessed 2023-07-16).

145. 日本バイオセーフティ学会 / *English Info.* https://jbsa-gakkai.jp/english_info.html (accessed 2023-07-16).
146. South Korea | Virtual Biosecurity Center. https://www.virtualbiosecuritycenter.org/governments/south-korea/ (accessed 2023-07-16).
147. Sijnesael, P. C. C.; van den Berg, L. M.; Bleijs, D. A.; Odinot, P.; de Hoog, C.; Jansen, M. W. J. C.; Kampert, E.; Rutjes, S. A.; Broekhuijsen, M.; Banus, S. Novel Dutch Self-Assessment Biosecurity Toolkit to Identify Biorisk Gaps and to Enhance Biorisk Awareness. *Front. Public Health* **2014**, *2*. doi: 10.3389/fpubh.2014.00197
148. An Efficient and Practical Approach to Biosecurity. https://www.biosecurity.dk/fileadmin/user_upload/PDF_FILER/Biosecurity_book/An_efficient_and_Practical_approach_to_Biosecurity_web1.pdf
149. CONICET | Buscador de Institutos y Recursos Humanos. https://www.conicet.gov.ar/new_scp/detalle.php?keywords=&id=20883&inst=yes&congresos=yes&detalles=yes&congr_id=7307076 (accessed 2023-07-17).
150. *ANLIS Malbrán.* Argentina.gob.ar. https://www.argentina.gob.ar/salud/anlis (accessed 2023-07-17).
151. *Argentina.gob.ar.* Argentina.gob.ar. https://www.argentina.gob.ar/ (accessed 2023-07-17).
152. World Health Organization. *Biorisk Management : Laboratory Biosecurity Guidance*; WHO/CDS/EPR/2006.6. World Health Organization, **2006**. https://apps.who.int/iris/handle/10665/69390 (accessed 2023-08-08).
153. 14:00-17:00. ISO 35001:2019. ISO. https://www.iso.org/standard/71293.html (accessed 2023-08-04).
154. Quarantelli, E. L. Emergencies, Disaster and Catastrophes Are Different Phenomena. *Environmental Science, Sociology* **2000**. https://www.semanticscholar.org/paper/Emergencies%2C-Disasters-and-Catastrophes-Are-Quarantelli/a4a1672d5bcb49c60f137bb3d5324bf71045123a
155. Caskey, S.; Gaudioso, J.; Salerno, R. *Biosecurity Risk Assessment Methodology (BioRAM) v. 2.0*; BIORAM v.2.0; Sandia National Lab. (SNL-NM), Albuquerque, NM, 2009. https://doi.org/10.11578/dc.20220414.20.
156. Zeng, X.; Jiang, H.; Yang, G.; Ou, Y.; Lu, S.; Jiang, J.; Lei, R.; Su, L. Regulation and Management of the Biosecurity for Synthetic Biology. *Synth. Syst. Biotechnol.* **2022**, *7*(2), 784–790. https://doi.org/10.1016/j.synbio.2022.03.005.
157. Brizee, S.; Passel, M. W. J. V.; Berg, L. M. V. D.; Feakes, D.; Izar, A.; Lin, K. T. B.; Podin, Y.; Yunus, Z.; Bleijs, D. A. Development of a Biosecurity Checklist for Laboratory Assessment and Monitoring. *Appl. Biosaf.* **2019**, *24* (2), 83–89. https://doi.org/10.1177/1535676019838077.
158. Clevestig, P. *Handbook of Applied Biosecurity for Life Science Laboratories.* SIPRI, **2009**.
159. *Laboratory Biosecurity Handbook.* Routledge & CRC Press (2007). https://www.routledge.com/Laboratory-Biosecurity-Handbook/Salerno-Gaudioso-Brodsky/p/book/9780849364754 (accessed 2023-07-24).
160. Biosecurity Central. https://biosecuritycentral.org/ (accessed 2023-08-27).
161. Prepare and Protect: Safer Behaviors in Laboratories and Clinical Containment Settings | Wiley. Wiley.com. https://www.wiley.com/en-us/Prepare+and+Protect%3A+Safer+Behaviors+in+Laboratories+and+Clinical+Containment+Settings-p-9781683670148 (accessed 2023-08-11).
162. Reynolds, A.; Lewis, D. Teams Solve Problems Faster When They're More Cognitively Diverse. *Harvard Business Review.* March 30, **2017**. https://hbr.org/2017/03/teams-solve-problems-faster-when-theyre-more-cognitively-diverse (accessed 2023-07-23).
163. What Are the Feds Doing about the Illegal Lab in Reedley? | YourCentralValley.com. https://www.yourcentralvalley.com/news/illegal-reedley-lab/what-are-the-feds-doing-about-the-illegal-lab-in-reedley/ (accessed 2023-11-06).

164. Biohacker Josiah Zayner is under Investigation –Vox. https://www.vox.com/future-perfect/2019/5/19/18629771/biohacking-josiah-zayner-genetic-engineering-crispr (accessed 2023-11-06).
165. Whitford, C. M.; Dymek, S.; Kerkhoff, D.; März, C.; Schmidt, O.; Edich, M.; Droste, J.; Pucker, B.; Rückert, C.; Kalinowski, J. Auxotrophy to Xeno-DNA: An Exploration of Combinatorial Mechanisms for a High-Fidelity Biosafety System for Synthetic Biology Applications. *J. Biol. Eng.* **2018**, *12*(1), 13. https://doi.org/10.1186/s13036-018-0105-8.
166. Beck, L. Laboratory Biosecurity. https://www.osti.gov/servlets/purl/1712988
167. Astuto-Gribble, L.; Caskey, S. Laboratory Biosafety and Biosecurity Risk Assessment Technical Guidance Document; SAND2014-15939R, 1171429, 533375; 2014; pp SAND2014–15939R, 1171429, 533375. https://doi.org/10.2172/1171429.
168. Novossiolova, T. A.; Whitby, S.; Dando, M.; Pearson, G. S. The Vital Importance of a Web of Prevention for Effective Biosafety and Biosecurity in the Twenty-First Century. *One Health Outlook* **2021**, *3*(1), 17. https://doi.org/10.1186/s42522-021-00049-4.
169. Yang, J.-Y.; Lee, S.-N.; Chang, S.-Y.; Ko, H.-J.; Ryu, S.; Kweon, M.-N. A Mouse Model of Shigellosis by Intraperitoneal Infection. *J. Infect. Dis.* **2014**, *209*(2), 203–215. https://doi.org/10.1093/infdis/jit399.
170. Government of Canada. Pathogen Safety Data Sheets: Infectious Substances – *Shigella* spp. https://www.canada.ca/en/public-health/services/laboratory-biosafety-biosecurity/pathogen-safety-data-sheets-risk-assessment/shigella.html (accessed 2023-07-16).
171. Lloyd, J. The Making of a Biosafety Officer. Issues in Science and Technology. https://issues.org/making-biosafety-officer-gillum/ (accessed 2023-09-24).
172. Biosecurity and Emerging Threats | Johns Hopkins | Bloomberg School of Public Health. https://publichealth.jhu.edu/departments/environmental-health-and-engineering/research-and-practice/research-areas/biosecurity-and-emerging-threats (accessed 2023-08-27).
173. PhD in Biodefense. Schar School of Policy and Government. https://schar.gmu.edu/programs/phd-programs/phd-biodefense (accessed 2023-08-27).
174. Vohun | Southeast Asia One Health University Network | Thailand. *SEAOHUN.org*. https://www.seaohun.org/vohun (accessed 2023-08-27).
175. House, T. W. FACT SHEET: CHIPS and Science Act Will Lower Costs, Create Jobs, Strengthen Supply Chains, and Counter China. *The White House*. https://www.whitehouse.gov/briefing-room/statements-releases/2022/08/09/fact-sheet-chips-and-science-act-will-lower-costs-create-jobs-strengthen-supply-chains-and-counter-china/ (accessed 2023-09-24).
176. Issues. Training More Biosafety Officers. Issues in Science and Technology. https://issues.org/training-more-biosafety-officers-gillum-forum/ (accessed 2023-09-24).
177. Committee on Space Research (COSPAR), COSPAR Book Series. COSPAR website. https://cosparhq.cnes.fr/cospar-book-series/ (accessed 2023-08-23).
178. Stuxnet. *Wikipedia*; 2024.

Index

Printed in the United States
by Baker & Taylor Publisher Services